Agricultural Plant Biochemistry

nipa books for everyone
PAPERBACKS

NIPA® GENX ELECTRONIC RESOURCES & SOLUTIONS P. LTD.
New Delhi-110 034

www.nipaers.com
Browse, Search, Read & Buy...
Print Books eChapters
eBooks Publishers
Forthcoming Subject Catalogues
New Titles
Online Resources on:
✓ Current Affairs
✓ Reasoning, Logic & Aptitude
✓ Competitive Examinations
✓ English Language Lab
✓ Writing & Pronunciation Tools
✓ Personality Development
✓ Online Programmes for Professionals
✓ Interview Preparation
Pay Using
PhonePe
Google Pay
paytm
PayPal
UPI
CC Avenue

Agricultural Plant Biochemistry

G. Nagaraj

Ex. Principal Scientist (Biochemistry)
Directorate of Oilseeds Research
Rajendranagar, Hyderabad - 500 030
Andhra Pradesh, India

nipa books for everyone PAPERBACKS

NIPA® GENX ELECTRONIC RESOURCES & SOLUTIONS P. LTD.

New Delhi-110 034

nipa® books for everyone
PAPERBACKS

NIPA® GENX ELECTRONIC
RESOURCES & SOLUTIONS P. LTD.

101,103, Vikas Surya Plaza, CU Block
L.S.C. Market, Pitam Pura, New Delhi-110 034
Ph : +91 11 27341616, 27341717, 27341718
E-mail: newindiapublishingagency@gmail.com
www: www.nipabooks.com

For customer assistance, please contact
Phone: + 91-11-27 34 17 17
Fax: + 91-11- 27 34 16 16
E-Mail: feedbacks@nipabooks.com

ISBN:978-81-96079-05-5

Composed and Designed by NIPA®.

This Book is Dedicated to
The Memory of My Beloved Parents

Smt. G. Muni Laxmamma & Sri. G. Krishna Swami Naidu

Foreword

Plant biochemistry is an important emerging field in the agricultural sciences. Basic knowledge of the chemistry and the biochemical mechanisms of the plant in synthesizing various components is essential for advancements needed in other areas of agriculture like plant breeding, plant protection, plant production, *etc*. In the recent past, biotechnology and biochemistry are moving hand in glove to solve many problems related to humans and other living beings. Plant genetic engineering is playing a major role in tackling the quantitative and qualitative aspects of food, fodder and fertilizer.

The challenges posed by the burgeoning population and dwindling land and water resources are umpteen on this earth. Plants being more efficient and environment friendly have a major role in solving many such problems. Plant production scientists need to understand the biochemical mechanisms for achieving the targets of food and fodder production. Source sink relationships need to be tackled beneficially to achieve higher productivity levels. Curtailing use of chemicals and conserving nature is another area that is engaging the attention of scientists.

This book namely, "Agricultural Plant Biochemistry" is aimed at providing good information to graduate and post-graduate students in agriculture and biology. I am sure, it will also serve as a reference book to researchers in plant breeding, agronomy, plant physiology and plant protection and will come in handy to solve many global problems by the present and future generations.

Dr. G. Nagaraj has served the Indian Council of Agricultural Research, Govt. of India for over 40 years till 2006 as a scientist in Plant Biochemistry. Post-retirement, he has been active as a Fellow of the Department of Science and Technology, Govt. of India, a teacher of Biochemistry, Engineering Chemistry, Biochemical Techniques and a Consultant in Bio-energy. He has published around 300 research articles as well as a few books on oilseeds and their quality. He has vast knowledge and experience as a Biochemist. I congratulate him in his efforts and wish that this book will have wide circulation in India and abroad.

(K.S. Varaprasad)
Project Director
Directorate of Oilseeds Research
Rajendranagar, Hyderabad-500 030, India

Preface

Plants have an important role on this globe. Being autotrophs they synthesize the most essential food components with the help of simple chemicals. They also have the ability to put together secondary nutrients, drugs and toxicants. Apart from sustaining their life, they help other forms of life including humans on this earth. They are also responsible for maintaining the climate and a congenial environment. Agriculture, which built up civilizations, has been the mainstay of human beings. Increasing population and changing needs of food with respect to quantity and quality has built up pressure to understand the biochemistry of plants and engineer them to suit to the requirements. This book is aimed at putting together the information on the various aspects of biochemistry of plants, their mechanism and utility. It has basic information and should be useful to beginners in agriculture, horticulture, biology and related disciplines.

As a researcher for over forty years in agriculture, I have been collecting information on plant biochemistry. I also depended on some website and teaching material. I tried to put together most such information. I am sure the book will be useful not only to beginners, but also to some specific researchers. It should be my endevour to improve the content and quality based on expert advises.

Hyderabad

May, 2014

G. Nagaraj

guttarla@hotmail.com

Contents

Chapter – 1

Introduction

All living organisms are basically made up of chemical components. The chemistry, composition and energy management of living organisms is what we call biochemistry. It tries to explain life processes at the molecular level. Although different organisms differ outwardly in their life processes, there are striking similarities in executing the different tasks. Genetic code, metabolic pathways, enzymes, coenzymes and even regulatory mechanisms are similar to a large extent in all the living organisms. Living organisms have certain extraordinary properties. They can grow, respond to stimuli and replicate themselves with high fidelity. All these activities are ultimately interpretable in chemical terms. The components that constitute organisms are all inanimate. The lifeless organic molecules with appropriate complexity and properties make a living being.

Biochemistry helps us understand how the inanimate molecules that constitute living organisms interact with each other to maintain life. The basic life processes or the biochemistry remains broadly the same in either the unicellular microorganism or the higher organisms such as the plants or the human beings. Life is nothing but thousands of ordered chemical reactions. In other words, chemistry is the logic of all biological phenomena.

Biomolecules

Living organisms are complicated and highly organized. Each component of the living organism has a specific purpose and function. All organisms have the ability to extract, transform and use energy from the environment.

Most of the components of organisms are organic compounds of carbon, hydrogen, oxygen or nitrogen. However, phosphorus and sulphur followed by some metals also are part of the living beings. Even the simplest organisms like bacteria contain many organic molecules. The number of compounds may range from 3000 to 5000 with 1000 nucleic acids. In a complex organism like the human being the number may go up to 50,000.

Carbon is the basis of life on this planet. It is a component of every organism. It is also the principal component of our daily diet, fuel and clothing. There are over 6 million carbon based compounds compared to around 250,000 inorganic compounds. Carbon, being present in the fourth group of the periodic table has intermediate electro-negativity. It can combine with elements on either side of it's fourth group and also with other carbon atoms resulting in a huge number of compounds. There are two reasons for the huge number of organic compounds. The first relates to the covalent bonding characteristics of the carbon and the second to the isomerism. Another characteristic is that carbon based organic compounds are highly stable.

Chemists up to 1828 thought that it was impossible to make organic (living) compounds. It was felt that "vital force" was required to make compounds present in living beings. A turning point took place by accident. Freidrich Wohler (1800-1882) on heating the inorganic compound ammonium cyanate, produced "urea" an organic compound present in blood and urine.

$$\underset{\text{Ammonium cyanate}}{\text{NH4 OCN}} \longrightarrow \underset{\text{Urea}}{\text{NH2--}\overset{}{\underset{\underset{\text{O}}{\|}}{\text{C}}}\text{--NH2}}$$

Organic and inorganic compounds differ in many of their characteristics. Some of them are summarised in the Table 1.

Table 1. Properties of organic and inorganic compounds.

Property	Organic	Inorganic
Bonding	Covalent	Ionic
Forces between molecules	Weak	Strong
Physical state	Gas, liquid/ solid	Usually solids
Solubility in water	Low	High
Rate of chemical reactions	Slow	Fast
Flammability	Flammable	Non-flammable

History of biochemistry

Only during the 17th and 18th centuries, important foundations were laid in many fields of biology. In the 18th and 19th centuries the link between biology and physical science was established. Important concepts relating to Biochemistry were proposed in the 19th century. Studies in Biochemistry have been so rapid that it is now a forerunner and the language of biology.

Scheele (1742-1786) isolated and studied citric acid, lactic acid and uric acid while Lavoisier (1743-1794) noticed that " burning" inside the body was similar to combustion and oxidation of organic molecules. He proposed that O_2 was consumed and CO_2 was released. Leibig (1803) showed that plants depended on inorganic molecules for their nutrition. Wohler (1802-1882) synthesized an organic compound, namely, urea. Prior to that only living beings were considered to synthesise organic compounds due to the presence of "vital force" in them.

Louis Pasteur, during 1857, did a great deal of work on fermentations and pointed out the importance of enzymes in this process. The breakthrough in enzyme research and hence, Biochemistry was made in 1897 by Edward Buckner when he extracted an enzyme from yeast cells in crude form which could ferment a sugar molecule into alcohol. This was almost a final blow to the theory of vitalism. Neuberg introduced the term Biochemistry in 1903. In 1926, James Sumner established the protein nature of enzymes. He was responsible for the isolation and crystallization of urease, which provided a breakthrough in studying the properties of specific enzymes. Arrhenius, Vant Hoff and Oswald worked on the concepts of osmotic pressure. Sorensen developed the concept of p^H.

In the 20th century, the growth was very fast in the field of Biochemistry. Structures of many compounds were formulated, for Ex: ATP structure was proposed by Fiske and Subbarow. The first metabolic pathway elucidated was the glycolytic pathway during the first half of the 20th century by Embden and Meyerhoff. Otto Warburg, Cori and Parnas also made very important contributions relating to the glycolytic pathway. Krebs established the citric acid and urea cycles during 1930-40. In 1940, Lipmann described the central role of ATP in biological systems.

The biochemistry of nucleic acids entered a phase of exponential growth after the establishment of the structure of DNA in 1953 by Watson and Crick followed by the discovery of DNA polymerase by Kornberg in 1956.

From 1960 onwards, Biochemistry plunged into an interdisciplinary phase sharing much in common with biology and molecular genetics. Frederick Sanger's contributions in the sequencing of protein in 1953 and nucleic acid in 1977 were responsible for further developments in the field of protein and nucleic acid research.

Living organisms exchange energy

Living organisms require a kind of energy *called free energy*, which can perform work at constant temperature and pressure. Cells in organisms give out energy in the form of heat, which is less useful. Living organisms are different from man made machines. They carry out the energy transformation reactions at normal or lower temperatures. The molecules (enzymes) which carry out these reactions are unstable and fragile.

All living cells are chemical engines that function at constant temperatures. The energy needs of all organisms are met from the solar energy. All living organisms are dependent on each other through exchange of energy and matter from or to the environment. Cells function as chemical engines because they possess enzymes which are biological catalysts functioning at ambient temperatures. Cells capture and store energy in the form of ATP (Adenosine Tri- Phosphate). It is the energy currency of all organisms.

Living cells are self regulating chemical engines operating on the principle of maximum economy. Genetic information is contained in a single molecule of DNA (Deoxy ribo- Nucleic Acid). A living cell is a self assembling, self adjusting, self perpetuating iso- thermal system of organic molecules, which extracts free energy and raw materials from it's environment. It carries out many consecutive reactions with the help of enzymes. These enzymes are also produced by the cell itself.

The cell maintains itself in a dynamic state and functions on the principle of maximum economy. It is self replicating over many generations. It is the aim of Biochemistry to understand the interactions of bio-molecules with each other and maintain the characteristics of living state.

Plant biochemistry

Plants are different from other living beings in that they are part of the environment and experience the stresses in and around them. Another important feature of plants is that they need only a few resources like light, water, CO_2 and minerals. Hence Plant Biochemistry is slightly different from Animal Biochemistry. Members of the plant kingdom are divided into two groups: autotrophic and heterotrophic. Autotrophic organisms can biosynthesise all organic molecules while heterotrophic organisms need a few complex organic molecules for which they depend on other organisms.

During germination the plant depends on the seed stores heterotophically. All green plants are autotrophic depending mostly on CO_2, H_2O, NO_3^-, SO_4^{2-} and PO_4^{3-}. Using these nutrients plants synthesise proteins, carbohydrates, lipids and nucleic acids etc. Lower forms of plant life like fungi are heterotrophic. They are not green and cannot tap solar energy. Hence they depend on autotrophic organisms to derive some of their energy.

The green plant possesses the unique machinery to convert light energy to chemical energy through a process called photosynthesis. Both autotrophs and heterotrophs have the ability to obtain energy by oxidation and utilize the same in the form of ATP. Bacteria vary with respect to autotrophy and heterotrophy. Generally the metabolic details of autotrophic bacteria are somewhat related to plant kingdom.

It is the aim of this book to present the important biochemicals that constitute the plants in general. How they are synthesized and inter-related also form an important aspect. The chemistry and properties of primary metabolites namely carbohydrates, lipids, amino acids, proteins etc. are initially discussed. Their inter-conversions through the biological catalysts, namely, enzymes along with their chemistry are also presented. Nucleic acids, structure, properties and their role in protein synthesis also find a place. Metabolic interactions relating to biosynthesis and catabolism in plants and the energy management are also examined in some detail. Other plant biochemicals like pigments, terpenoids, steroids, alkaloids, phytohormones etc. and their properties are also discussed.

Nitrogen metabolism and its importance in crop production is presented in some detail. Plant cell, its organisation and utility finds a brief mention. The role of different biochemicals and their interactions in Agriculture has been mentioned, wherever possible.

Chapter – 2

Carbohydrates

Carbohydrates are the most abundant biomolecules on earth. They occur in nature as glucose, starch and cellulose etc. Each year, photosynthesis converts more than 100 billion metric tonnes of carbon dioxide and water into starch, cellulose and other plant products. Carbohydrates are polyhydroxy aldehydes or ketones and their derivatives. They provide 4k calories energy per gram. They promote utilization of lipids and reduce wastage of proteins. They are readily available in roots, tubers, cereal grains, sugarcane and sugar beet.

Carbohydrates have the general formula $C_n(H2O)_n$, indicting that they are the hydrates of carbon. Some carbohydrates also contain nitrogen, phosphorous or sulfur. They are characterized by the number of carbon atoms they contain. If n=3, they are trioses, n=4 tetroses, n=5 pentoses, n=6 hexoses and n=7 heptoses. The most common sugars in plants are the hexoses. Two or more simple carbohydrates condense together to form polymer sugars like sucrose, starch and cellulose etc. The basic condensing bond is glycosidic bond.

A carbohydrate is also called saccharide, meaning sugar. Sugars are synthesized by plants through a series of reactions from carbon dioxide in the presence of sunlight and a green pigment called chlorophyll by a process named photosynthesis. Sugars serve as sources of energy to all the living beings. Sugars are broken down through a process called respiration to release energy.

$$6CO2+6H2O \underset{\text{respiration}}{\overset{\text{photosynthesis}}{\rightleftharpoons}} C6H12O6+6O2$$

Functions of carbohydrates

1. Carbohydrates such as sugar (sucrose) and starch are dietary staple food in most parts of the world while cellulose is the feed of herbivorous animals. The oxidation of glucose is the energy yielding pathway in most cells of living beings.
2. They are the reserve or storage forms of energy in plants (starch, inulin) and in animals (glycogen).
3. The insoluble carbohydrate polymers serve as structural and protective elements in the cell walls of bacteria and plants and in connective tissue of the animals i.e., they give structural rigidity. Ex: cellulose in cell wall, chitin in insects and mucopolysaccharides in bacteria.
4. They are important components of nucleic acids, coenzymes and flavoproteins.

Classification of carbohydrates

Carbohydrates vary greatly in size ranging from the simple glyceraldehyde molecule to the plymeric starch molecule. The carbohydrates have been classified based on the degree of polymerization into mono, di, oligo, and polysaccharides. Saccharide in Greek means sugar.

Monosaccharides: These are the simple sugars which cannot be further broken down. They consist of a single polyhydroxy aldehyde or ketone unit. Depending on the number of carbon atoms, there are monosaccharides like trioses, tetroses, pentoses, hexoses and heptoses. Ex. glyceraldehydes(glycerose), erythrose, ribose, glucose, galactose, fructose etc. Glucose is the major source of energy in animals. Fructose is the fruit sugar.

Disaccharides and Oligosaccharides: They are composed of two (di) or a few (oligo) monosaccharides joined together by covalent bonds. The most abundant are the disaccharides like sucrose, maltose etc. A disaccharide can be split into two mono-saccharide units. Sucrose on hydrolysis yields glucose and fructose while maltose yields two glucose units. Oligosaccharides occur as side chains of plypeptides in glycoproteins. For example, the trisaccharide raffinose yields on hydrolysis galactose, glucose and fructose while stacchyose yields two galactose units, one glucose unit and one fructose unit.

Polysaccharides: They are composed of long chains having hundreds or thousands of monosaccharide units. Cellulose and starch are the most popular polysaccharides that occur in plant material. Both of them have recurring units of glucose. They are thus polymers of glucose.

Conjugated polysaccharides: In addition to carbohydrates, these polysaccharides contain proteins (glycoproteins) and lipids (glycolipids) also.

Table. 1. Classification of carbohydrates

Major types	1. Simple carbohydrates Ex. monosaccharides 2. Complex carbohydrates. Ex. di- oligo- and polysaccharides
Carbon number	1. Tetrose (C_4) 2.pentose (C_5) 3.hexose (C_6) 4.heptose (C_7)
Functional group	1. Aldoses : sugars with –CHO group 2. Ketoses : sugars with -C=O group
Reactivity	1. Reducing sugars: sugars oxidized by Tollens' reagent 2. Non-reducing sugars: sugars not oxidized by Tollens' reagent

Numbering carbons in sugars: The carbons are numbered starting from the aldehyde group as number 1. In the case of ketoses, the carbon of ketone group is the carbon number 2.

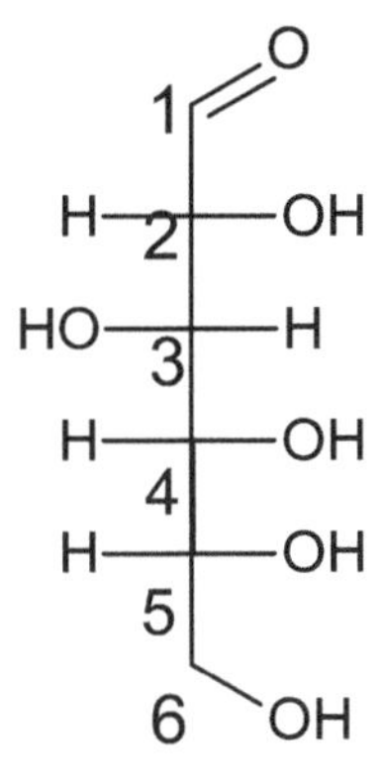

D- Glucose

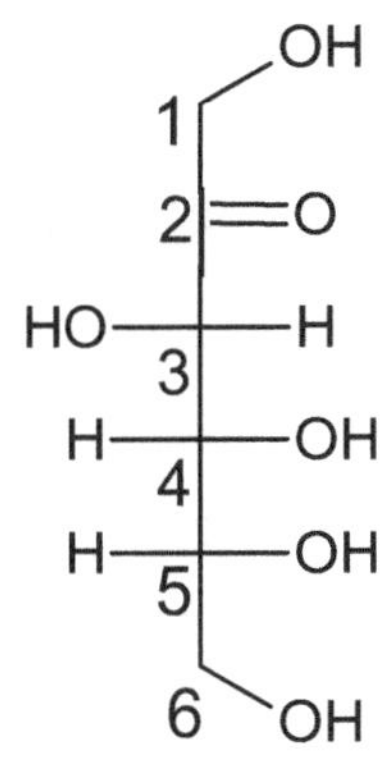

O- Fructose

Monosaccharides

They cannot be further hydrolysed under normal conditions of hydrolysis. They are further classified based on the number of carbon atoms present in the monosaccharide into trioses (3C), tetroses (4C), pentoses (5C), hexoses (6C) and heptoses (7C). Based on the functional group, they are classified as aldoses and ketoses depending on whether they have aldehyde or ketone as functional group.

Aldoses and ketoses: Monosaccharides are colorless, crystalline solids that are perfectly soluble in water and have sweet taste. Aldose is a compound which has an aldehyde (-CHO) at the end of it's chain while ketose has a ketone (C=O) group on the second carbon atom.

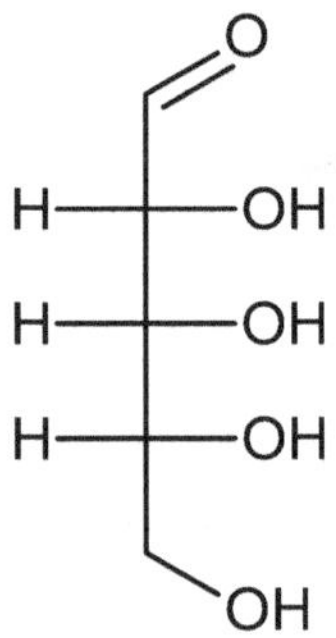

D- Ribose

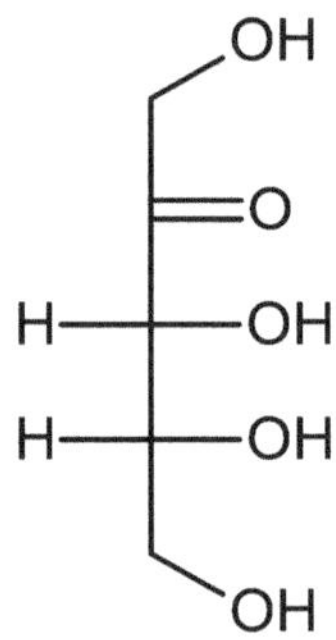

D- Ribulose

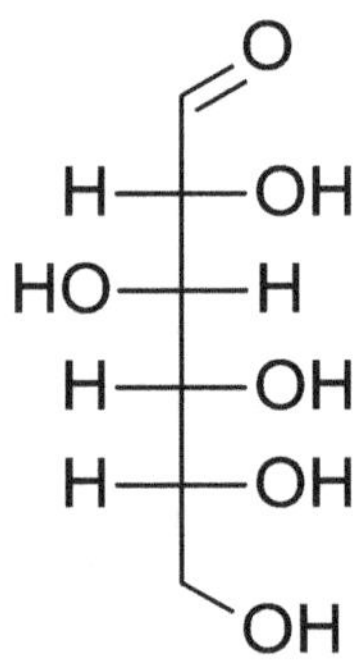

D- Glucose

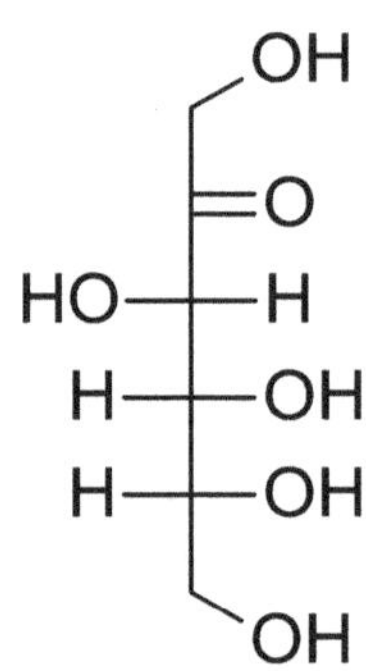

D- Fructose

The simplest monosaccharides are the 3 carbon trioses, namely, glyceraldehyde and dihydroxy acetone. Those having 4,5,6,and 7 carbon atoms are called tetroses, pentoses, hexoses and heptoses. Each of these exists in two series, namely, as aldoses and ketoses, ex.aldotetrose and ketotetrose. D-glucose (aldohexose) and D-fructose (ketohexose) are the most abundant. Glucose occurs in ripe fruits, flowers, honey, beetroot etc. Fructose or levulose is present in honey and fruits. It is the building block of the polysaccharide, inulin and sucrose. Galactose is the component of milk and agar agar. Mannose is a component of alginic acid. The pentose sugars arabinose and xylose occur as part of plant gums. D-ribose (aldopentose) and 2-deoxy D-ribose are the components of nucleic acids.

Examples of commonly occurring aldoses

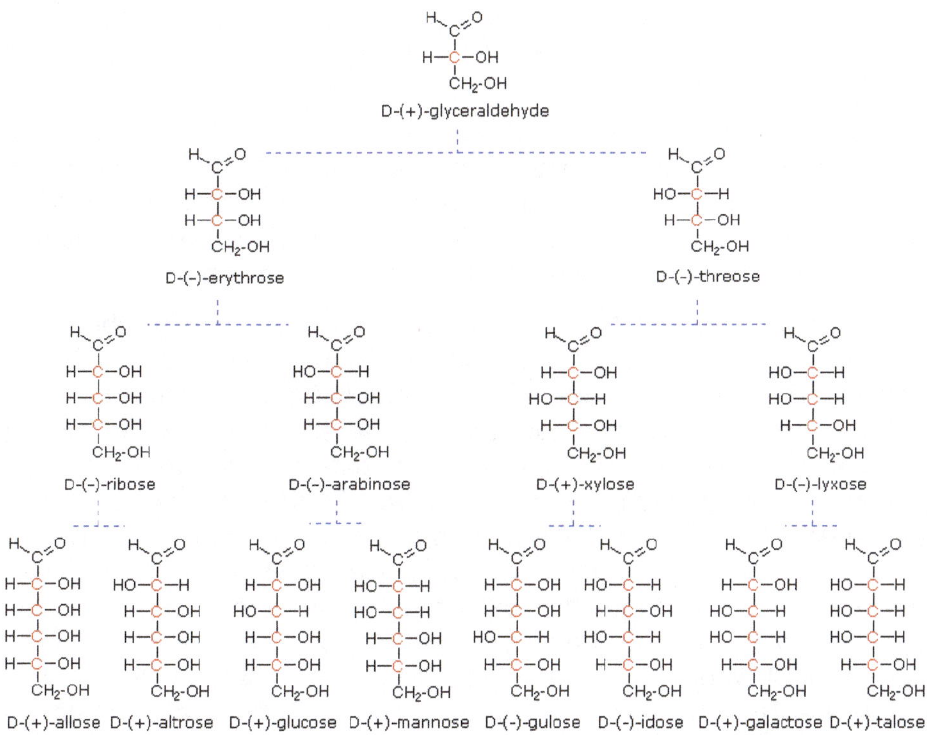

If a monosaccharide has a carbonyl function on one of the inner atoms of the carbon chain it is classified as a ketose. Dihydroxyacetone may not be a sugar, but it is included as the ketose analog of glyceraldehyde. The carbonyl group is commonly found at C-2. D-Fructose is the sweetest of the common natural sugars. Although the ketoses are distinct isomers of the aldose monosaccharides, the chemistry of both classes is linked due to their facile interconversion in the presence of acid or base catalysts. This interconversion, and the corresponding epimerization at sites alpha to the carbonyl functions, occurs by way of an enediol tautomeric intermediate.

Ketose Examples

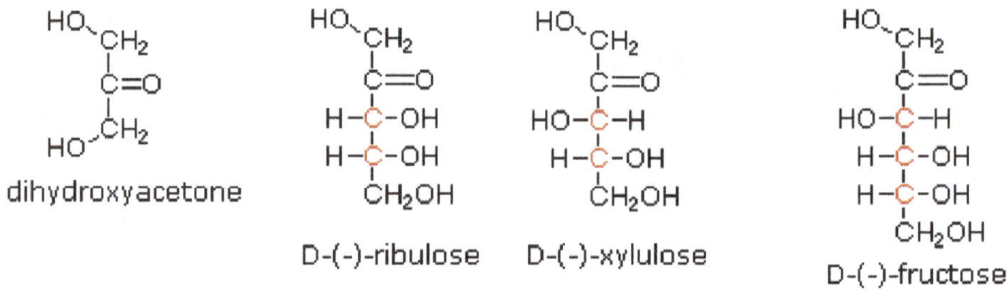

Isomerism

The term isomer was coined by J.J.Berzelius in 1827. It was used to differentiate compounds with the same structural formulae. Isomerisms are of different types depending on the arrangement of the carbon atoms and the various functional groups. They are stereoisomerism, structural isomerism, optical isomerism and functional isomerism.

a) *Stereoisomerism:* Most of the monosaccharides contain the same number of atoms and the same kinds of groups, yet they are definitely distinct substances. For example, the formula C6H12O6 represents 16 different simple sugars, all possessing the structure CH2OH CHOH CHOH CHOH CHOH CHO. This is due to different arrangement of the constituent groups of the molecule in space. This phenomenon is called as stereoisomerism and these sugars are called as stereoisomers. For example; Glucose, mannose and galactose are stereoisomers. When there are several asymmetric carbon atoms in a chain molecule and the end groups are not identical, the number of stereoisomers possible is equal to 2^n where n is the number of asymmetric carbon atoms. Thus there are 16 stereoisomers possible corresponding to the formula CH2OH CHOH CHOH CHOH CHOH CHO, which contains 4 asymmetric carbons ($2^4 = 16$).

Monosaccharides belong to D or L series depending on the position of OH group on the penultimate carbon atom. If OH is towards the right side of the penultimate carbon atom it is called as D sugar and if OH is towards the left side of the penultimate carbon atom it is called as L sugar. Glyceraldehyde, the simplest sugar is used as a reference compound for representing D and L forms of sugars. D and L forms of sugars which are non-super imposable mirror images of each other are called *enantiomers.*

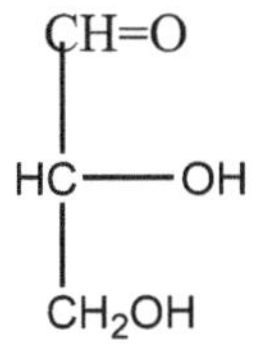

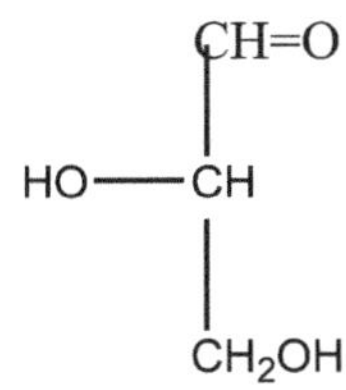

D-GLYCERALDEHYDE L-GLYCERALDEHYDE

In nature, D-sugars are more widely distributed than L-sugars. The stereoisomers which are not mirror images of each other are called *Diastereomers.* Eg: 1) D-Erythrose and D- Threose 2) D- Glucose, D-Mannose and D-Galactose. Among the diastereoisomers, those which differ in configuration at a single carbon atom are called *epimers.* Mannose is an epimer of glucose at 2nd carbon atom whereas galactose is an epimer of glucose at 4th carbon atom whereas galactose and mannose bear no epimeric relationship.

b) Structural isomerism: Some compounds have the same molecular formula but different structural formulae. For example Glucose, galactose and mannose have the same molecular formula but different structures and hence they are called structural isomers.

c) Functional isomerism: Glucose and fructose have the same molecular formula but glucose is an aldose while fructose is a ketose. This kind of isomerism is called functional isomerism.

d) Optical isomerism: Carbohydrates exhibit another kind of isomerism called optical isomerism. It is shown by the compounds having an asymmetric carbon. They have the same molecular and structural formulae but differ in their behavior towards plane polarized light. An optical isomer rotating the plane of polarized light toward right is called dextrorotatory 'D' (+) while that rotating the plane of polarized light toward left is called levorotatory 'L' (-).

Optical activity is the ability of a substance to rotate plane polarized light either to the right or left. Mutarotation is a gradual change of specific rotation of any optically active substance (sugar) having CHO or C=O group. When alpha D-glucose is dissolved in water, it's specific rotation gradually changes with time and reaches a stable value of 52.7^0. Alpha D-glucose when freshly dissolved has a specific rotation of $+112^0$ while beta D-glucose has +19. When the isomers are in equilibrium the value is 52.7^0.

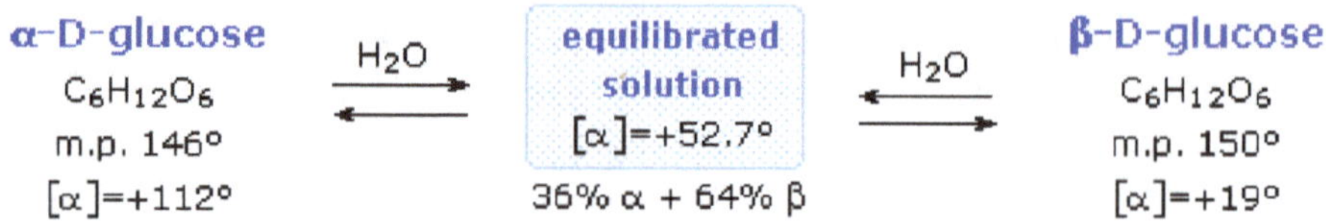

Tautomerism: A carbonyl compound with a 'H' atom on the α–carbon exists in equilibrium with an isomeric form in which 'H' atom is shifted to the carbonyl oxygen atom.

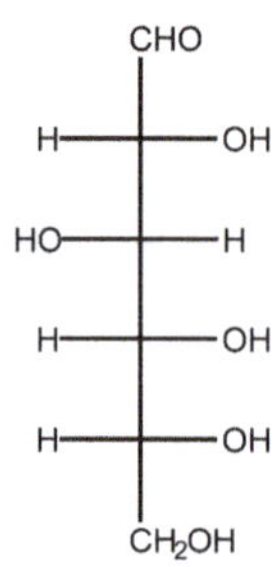

D-GLUCOSE

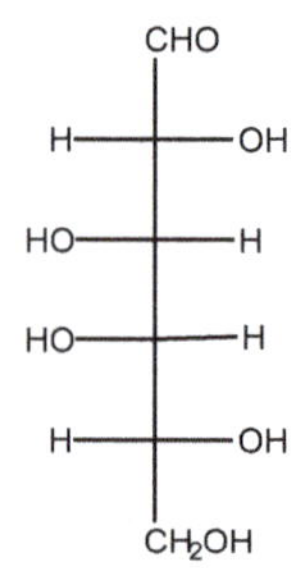

D-GALACTOSE

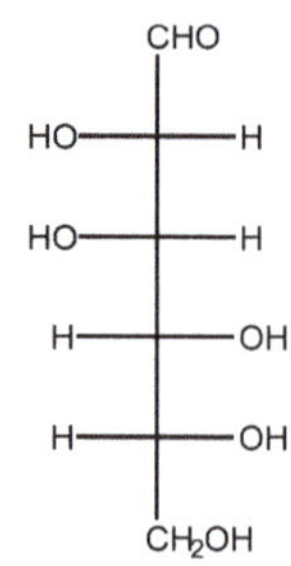

D-MANNOSE

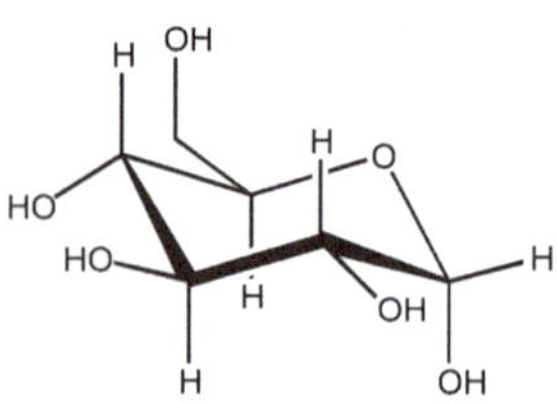

D-GLUCOSE -ALPHA PYRANOSE

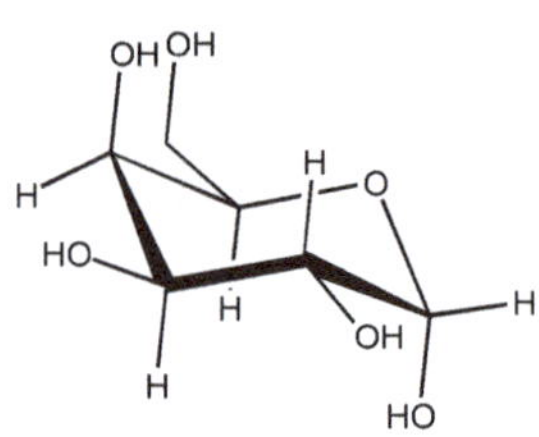

D-GALACTOE-ALPHA PYRRANOSE

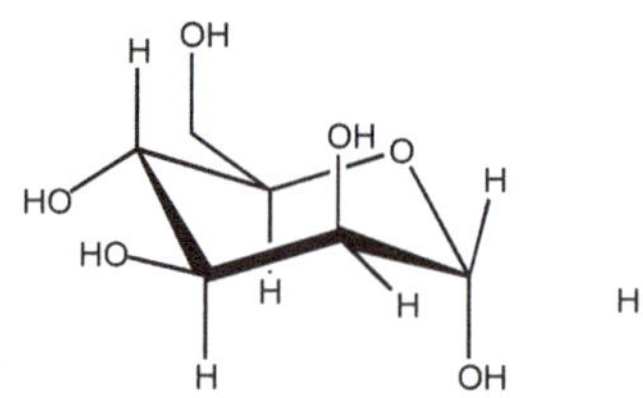

D-MANNOSE-AL
PHA PYRANOSE

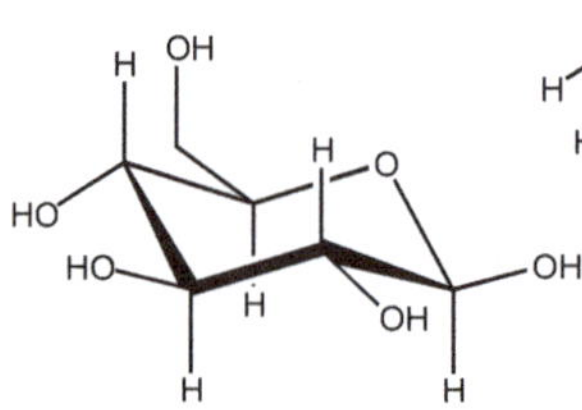

D-GLUCOSE- BETA PYRANOSE

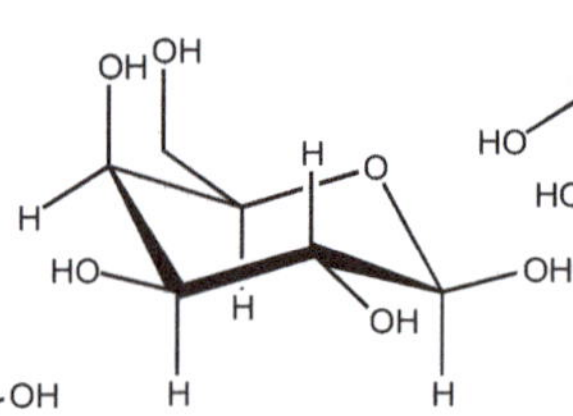

D-GALACTOSE-BETA PYRANOSE

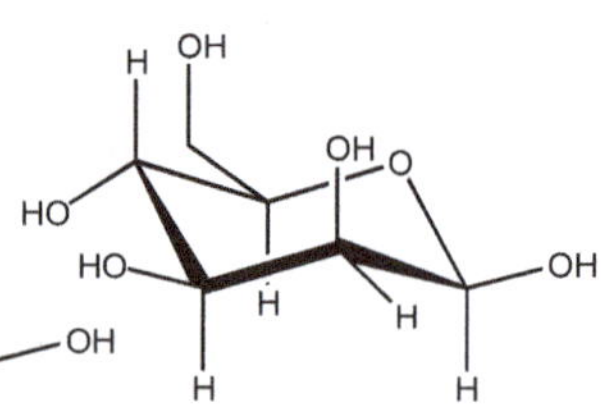

D-MANNOSE-BETA
PYRANOSE

The general structure of glucose and many other aldohexoses has been established with the help of simple chemical reactions. The following diagram shows some of the reactions. Based on the products obtained, the formula and structure of glucose has been deduced.

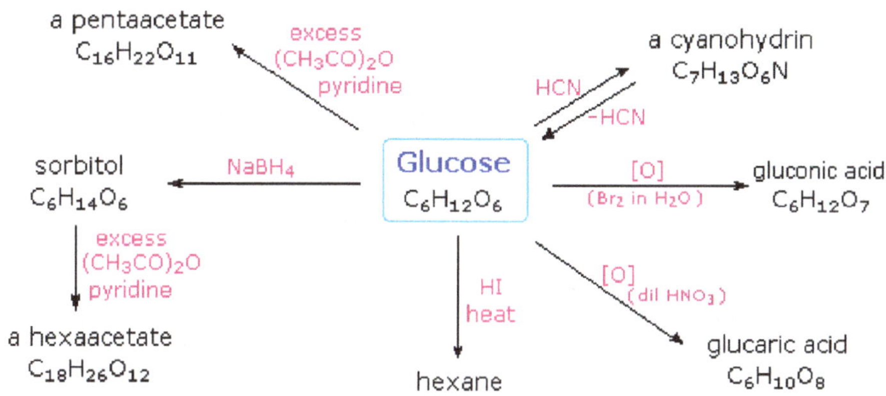

The cyclic structure for glucose: Glucose neither responds to the Schiffs test nor forms a suitable addition compound with sodium bisulphate. A freshly crystallized glucose has a specific rotation of +112^0. After storage it changes to +52.7^0. This change in rotation, namely, mutarotation suggests a change in structure. Normally an aldehyde reacts with two molecules of methanol to form an acetal. But the glucose molecule reacts with only one molecule of methanol. The methyl glucoside formed can be resolved into two forms. This information suggests the existence of a ring structure. Complete methylation of glucose and oxidation yields trimethyl derivative of glutaric acid, which shows that the oxygen is attached to the carbon 5 in the molecule.

Ring Structures: The aldehyde or ketone group of a monosaccharide can react with a hydroxyl group to form a covalent bond. Formally, the reaction between an aldehyde and the hydroxyl group of a sugar (an alcohol) creates a hemiacetal structure whereas a ketone reacts with hydroxyl group of a sugar (alcohol) to form a hemiketal structure. For tetroses and larger sugars, the reaction can take place within the same molecule so that the straight chain form of the sugar cyclizes. The following figure shows the cyclization of D-glucose to form a six-carbon ring.

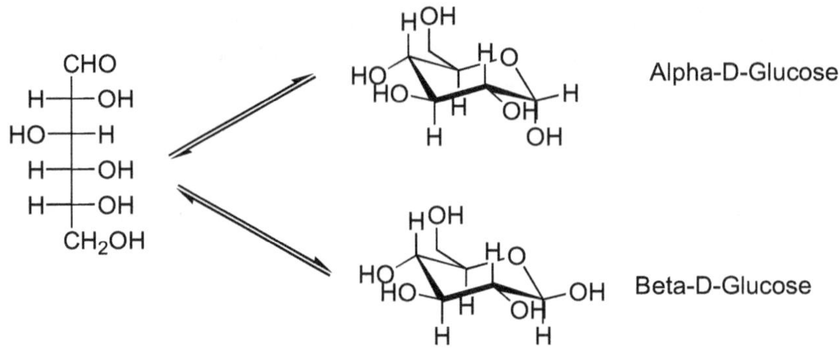

Cyclization of glucose

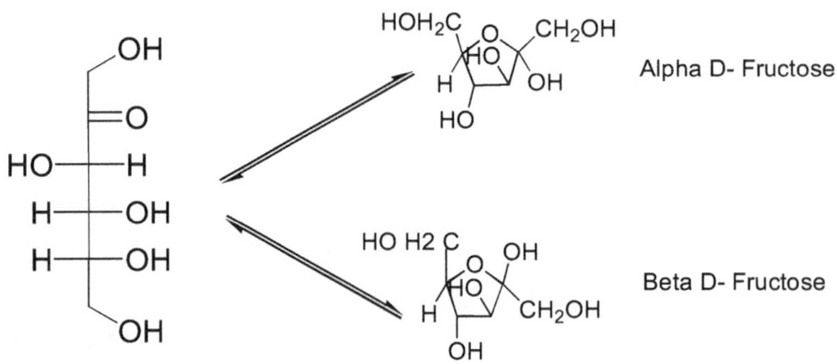

Cyclization of D-Fructose

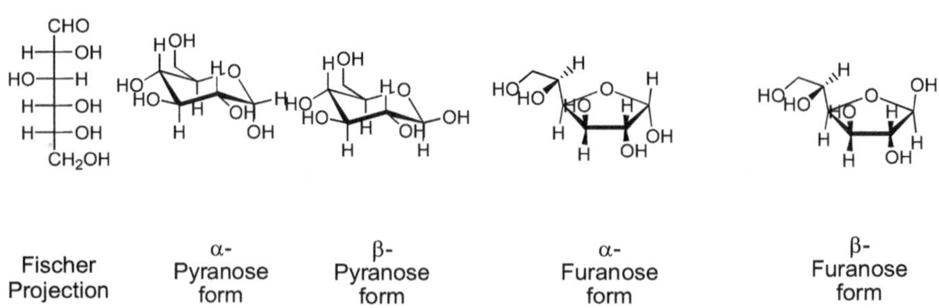

A new asymmetric center is formed during cyclization of an aldehyde at C-1.

1. Two isomers of glucose exists as α– D-Glucose in which the OH group at C- 1 lies below the plane of the ring and β-D-Glucose in which the OH group at C-1 lies above the plane of ring. The C-1 carbon is

called *anomeric* carbon atom and the α and β forms are called anomers. In aqueous solution the alpha and beta forms are rapidly inter convertible via the open chain structure to give an equilibrium mixture and this is termed as *mutarotation*. Because of the structural similarity to the ring compound called pyran, the six membered ring structures of hexoses are called pyranoses.

Five membered sugars such as D-ribose and D-deoxyribose and six carbon ketose sugars such as D-fructose, form rings called furanoses as their structures are similar to the furan ring. Even glucose can exist, although not in nature, as a five membered ring. This ring containing four carbons and one oxygen atom is referred to as furanose as distinct from the normal pyranose form. Fructose also can exist in the pyranose and furanose forms. However, fructose occurs more widely in the furanose form.

Again the furanoses can exist both in alpha and beta forms except where the nomenclature refers to the hydroxyl group attached to C-2 which is the anomeric carbon atom. The pyranose ring of a six-carbon aldose sugar can exist in either a boat or a chair configuration. The substituents attached to the ring carbons that extend parallel to the symmetry axis are said to be axial whilst those that extend outward from this axis are said to be equatorial. In the boat form, there is considerable steric hindrance between the various groups attached to the carbon atoms of the ring and therefore this form is less favorable energetically. Hence the chair form predominates, as shown for β-D-glucose where all the axial positions are occupied by hydrogen atoms.

There are several naturally occuring sugars containing five carbon atoms including D-ribose, D-ribulose, D-xylose and L- arabinose. Ribose and 2-deoxy ribose are the components of nucleic acids.

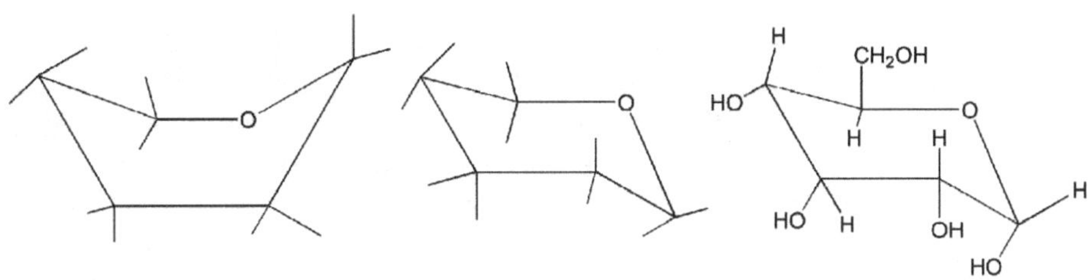

BOAT CHAIR ALPHA-D-GLUCOPYRANOSE

Oligosaccharides

The monosaccharides condense with each other through glycosidic linkage to form oligosaccharides. The oligosaccharides are further classified depending upon the number of monosaccharide units present. Two monosaccharides combine to form a disaccharide while three combine to form a trisaccharide and four combine to form a tetra saccharide. The commonly found oligosaccharides are sucrose, maltose, lactose, raffinose and stacchyose. Sucrose is the most abundant in sugarcane, sugarbeet and maple syrup. It is the most popular table sugar. A polysaccharide contains many monosaccharides. Starch and cellulose are the most important polysaccharides of plants.

Disaccharides: A disaccharide is composed of two monosaccharides. The hydroxyl group on the anomeric carbon atom of one monosaccharide can react with the hydroxyl group of a second monosaccharide to form a disaccharide. The covalent bond formed is called a glycosidic bond. Two monosaccharides combine to form a disaccharide by elimination of one molecule of water. The disaccharide is characterized by the nature of the two sugars which join together and also by the linkage. Two molecules of glucose can join together with condensation between the C_1 of one molecule and C_4 of the other. This can result in two different types of molecules since the first molecule may have an alpha or beta configuration. Two molecules of glucose condensing in the alpha position give rise to 1:4 alpha linked disaccharide called maltose. If C_1 in the first glucose molecule is in the beta- position, then the disaccharide cellobiose is produced with a 1:4 beta link.

a) *Lactose*: It is a disaccharide formed between the anorexic carbon C-1 of β-D-galactose and C-4 of α-D glucose. Since the anomeric carbon of galactose molecule is involved in the bond and is in the beta-configuration, this is called β (1:4) bond which can be abbreviated as β 1:4. It is the component of milk and is called milk sugar. It is present up to 8% in human milk and up to 5% in cow's milk.

b) *Maltose*: It is a disaccharide formed between the C-1 and C-4 positions of two α-D glucose units. However, here the configuration of the anomeric carbon atom involved is the alpha form and hence the bond is called an α (1:4) bond or abbreviated as α 1:4. For lactose and maltose, one of the anomeric carbons has been used to form the bond, leaving the second anomeric carbon free. Thus both lactose and maltose have a reducing end. Hence they have reducing property and are called reducing sugars.

c) Sucrose: It is a disaccharide formed by glycosidic bond formation between the anomeric C-1 of α-D glucose and the anomeric C-2 of β-D fructose so that sucrose lacks a free reducing group. Thus sucrose is a non reducing disaccharide. It is formed by condensation of glucose and fructose. It occurs upto 20% in sugar cane and upto 15% in sugar beet. It is the table sugar utilized up to 45 kg per annum by each person on this earth.

Tri- and tetra-saccharides: Three monosaccharide units condense with each other to form a trisaccharide. Eg: Raffinose is formed by condensation of Galactose, Glucose and Fructose. It is alpha D-galactopyranosyl (1-6)-O-D-glucopyranosyl (1-2) alpha-D- fructofuranoside. To put it simply raffinose is a sucrose molecule to which a galactose unit is attached.

Stacchyose is a tetrasaccharide containing two galactose units, one glucose unit and one fructose unit. It is D-galactopyranosyl (1-6)- O-D-galactopyranosyl (1-6)-O-alpha D-glucopyranosyl (1-2) beta D-fructofuranoside. Stacchyose is again a sucrose molecule to which two galactose units are attached. Also it can be considered as raffinose molecule to which an extra galactose unit is attached. Raffinose and stacchyose occur in oilseeds and are responsible for their bad taste and also the flatulence properties .

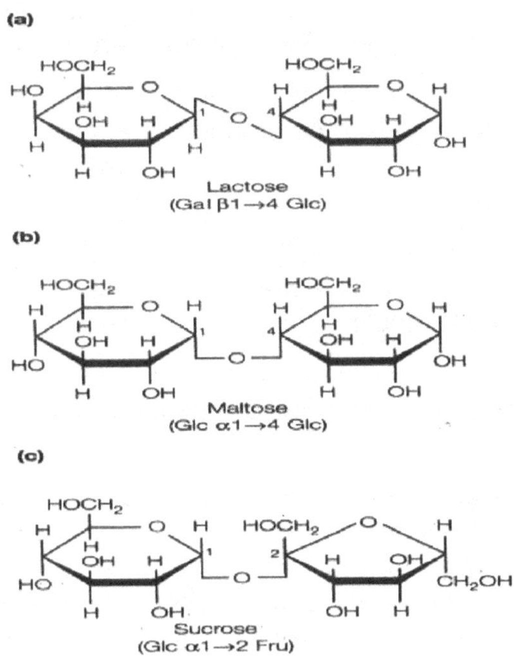

Polysaccharides

Many monosaccharide units condense to form polysaccharides through glysosidic linkages. To put it simply, a polysaccharide is the polymer of monosaccharides. They are the more widely occurring saccharides. Polysaccharides are high molecular weight compounds composed of a large number of monosaccharides. They are sometimes called glycans. The most abundantly available polysaccharides are starch, dextrins, glycogen, cellulose, hemicelluloses, pentosans and pectic substances. They are classified based on the function, nature of branching and the repeating units.

1. Functional classification

a. *Structural polysaccharide:* Polysaccharides belonging to this class help in maintaining the cell structure. Ex: Cellulose, chitin, hemicellulose, pectin etc.

b. *Storage polysaccharide:* Polysaccharides belonging to this class help in storing carbohydrate material in the cell. Ex: Starch, glycogen and inulin.

2. Nature of branching

a. *Linear:* Polysaccharides belonging to this class have a linear glycosidic bonding only. Ex: Cellulose, chitin and amylose

b. Branched: Polysaccharides belonging to this class have a branched glycosidic bonding. Ex: Amylopectin and glycogen.

3. Repeating Unit

a. *Homopolysaccharide:* Polysaccharides belonging to this class contain the same basic repeating monosaccharide unit. Ex:Starch, glycogen, chitin, inulin

b. *Heteropolysaccharides:* Polysaccharides belonging to this class contain more than one basic repeating unit. Ex: Hemicellulose, pectin.

Polysaccharides are long chains containing many simple sugar units. Depending on the polysaccharide, the chains may be linear or branched. In plants, the storage form of glucose is the polysaccharide called as starch whereas in animals excess glucose is stored as a large branched polysaccharide called glycogen. These polysaccharides serve as nutritional reserves and when required they are broken down and the monosaccharide products are metabolized to yield energy. In contrast, cellulose is present in cell walls and behaves as a structural polysaccharide.

Starch: Starch exists in plants as insoluble starch granules in the cytoplasm. It occurs in cereal grains, pulses, tubers, bulbs, and fruits ranging from a few per cent to 75%. Each starch molecule contains a mixture of two polysaccharide forms, amylose and amylopectin. Amylose is unbranched polymer of glucose residues joined in α 1:4 linkages. Amylopectin is the branched form in which most of the glucose residues are joined in α 1:4 linkages but additional α 1:6 bonds occur at every 25-30 residues repeating the branch points. Starch does not exist in one plane, but forms a spiral. Amylose is easily soluble in water while amylopectin is not easily soluble in water.

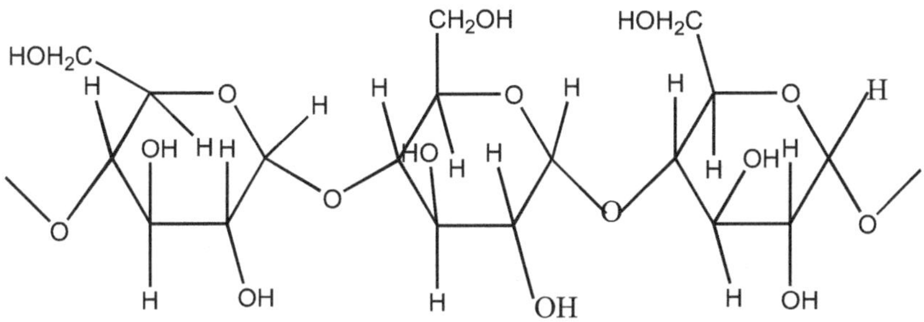

AMYLOSE

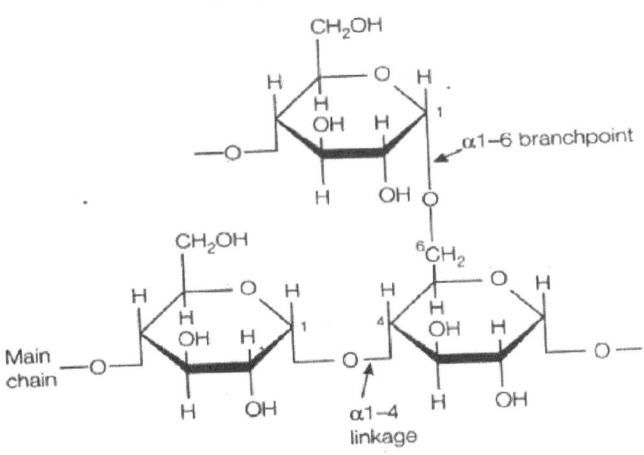

AMYLOPECTIN

Starch is the principal seed reserve of most plants, particularly cereals. In the living plant it is often laid down in winter, in the xylem parenchyma of woody plants, or in the parenchyma of storage organs such as potato tuber and then in the spring hydrolyzed and utilized.

Most animals, including humans, depend on these plant starches for their nourishment. The structure of starch is more complex than that of cellulose. The intact granules are insoluble in cold water, but grinding or swelling them in warm water causes them to burst. The released starch consists of two fractions. About 20% is a water soluble material called amylose. Molecules of amylose are linear chains of several thousand glucose units joined by alpha C-1 to C-4 glycoside bonds. Amylose solutions are actually dispersions of hydrated helical micelles. The majority of the starch is a much higher molecular weight substance, consisting of nearly a million glucose units, and called amylopectin. Molecules of amylopectin are branched networks built from C-1 to C-4 and C-1 to C-6 glycoside links, and are essentially water insoluble.

Glycogen: Glycogen molecule consists of glucose units which are linked in long chains by α 1:4 bonds. For every 10 units or so, the chain is branched by the formation of α 1:6 glycosidic bond. The glycogen chain terminates in a non reducing end with a free 4'-OH group. Since the enzyme that degrades glycogen catalyzes the removal of glycosyl units from non reducing end of glycogen chain, the numerous branches, each with a non reducing end, greatly increase the accessibility of the polysachharide to degradation. The α1:6 branches are removed by debranching enzyme. This is the storage polysaccharide of animals and hence is called animal starch. It is similar to amylopectin, but more branched than amylopectin.

Dextran is a glucose polymer where the glucose residues are mainly linked by the α 1:6 bonds. A few branches also occur which are formed by α 1:2, α 1:3 or α 1:4 bonds depending on the bacterial or yeast species that is the source of dextran. These are formed when starch is subjected to dry heat. They are intermediate between starches and simple sugars.

Cellulose: Cellulose is an unbranched polysaccharide of glucose units linked by β 1:4 bonds. This structure is not in a spiral, but lies in a plane. The glucose residues in cellulose are arranged as straight fibrils. In plant cell walls, the cellulose fibrils are embedded in a matrix of other polysachharides. In wood, the matrix also contains lignin, a complex polymer of phenolic residues.

Over half of the total organic carbon in the earth's biosphere is in the form of cellulose. Cotton fibers are essentially pure cellulose, and the wood of bushes and trees is about 50% cellulose. As a polymer of glucose, cellulose has the formula $(C_6H_{10}O_5)_n$ where n ranges from 500 to 5,000, depending on the source of the polymer. The beta-glycoside bonds in cellulose permit the chains to stretch out, and this conformation is stabilized by intra-molecular hydrogen bonds. A parallel orientation of adjacent chains is also favored by intermolecular hydrogen bonds. Although an individual hydrogen bond is relatively weak, many such bonds acting together impart great stability to certain conformations of large molecules.

Mammals including humans, lack the enzymes capable of digesting the β1:4 linkages of cellulose and so cannot digest plant cell walls. It acts as a roughage and dietary fibre in humans. It can be broken down to glucose by some enzymes and microorganisms. This is the way cellulose is hydrolyzed in the rumen of animals to provide them with sugar. Cellulose makes up more than 25% of the cell wall. It accounts for more than 50% of the world's supply of carbon.

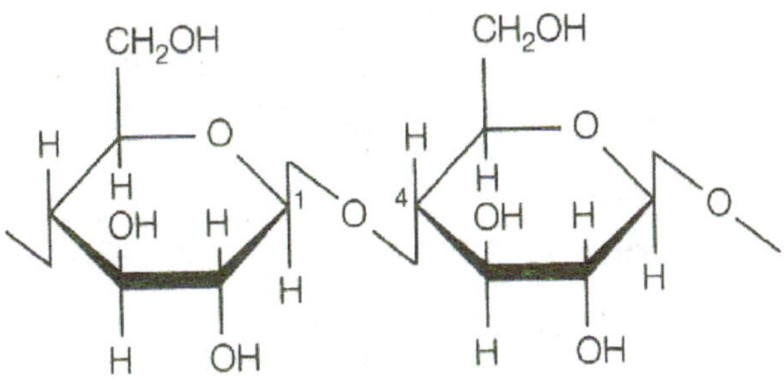

Cellulose

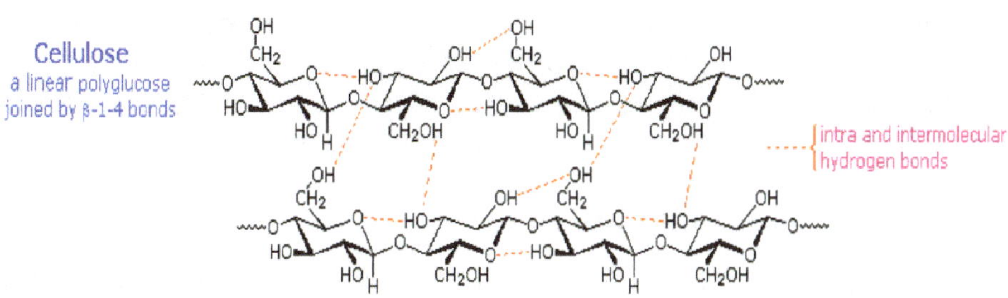

Hemicelluloses: Hemicelluloses are a poorly defined group of compounds. They are insoluble in water but unlike cellulose are soluble in alkali. Along with cellulose, hemicelluloses are also useful as dietary fibre. Pentoses, hexuronic acid and sometimes L-rhamnose are the constituents of hemicelluloses.

Pectic substances: These occur as components of plant cell walls and in the middle lamellae. They serve to hold the cells together. They are mixtures of polysaccharides formed from galactose, arabinose and galacturonic acid, the last being predominant. The carboxyl groups of galacturonic acid are partly esterified with methanol. Pectin forms a gel with sugar and acid under suitable conditions. Commercial pectin is extracted from citrus peels.

Gums: Gums are hydrophilic substances that form a viscous solution with hot or cold water. Most gums are polysaccharides like starch, pectic substances and derivatives of starch and cellulose. Natural gums are seed gums (Ex. guar gum), plant exudates (Ex. gum arabic) and seaweed gums (Ex. agar gel).

Conjugated polysaccharides: Besides occurring in free state, the carbohydrates occur in nature in conjugation with other biomolecules like lipids and proteins to form glycolipids and glycoproteins. Mucopolysaccharides are glycoproteins characterized by the presence of amino sugars like glucosamine, galactosamine, Ex. Hyaluronic acid and Heparin.

Important Reactions and properties of sugars

Most sugars are colourless crystalline solids. Mono- and disaccharides are soluble in water and are sweetish in taste. The sweetness of sugars are as follows: If sucrose=100, glucose=70, fructose= 170, maltose=30 and lactose=16. They also exhibit optical activity and mutarotation. Polysaccharides like starch form colloidal solutions on heating. Others like cellulose are insoluble in water. The glycosidic bonds of di- and polysaccharides are readily hydrolysed by acid but resist cleavage by bases. They yield monosaccharides on hydrolysis. Polysaccharides do not exhibit the properties of either ketones or aldehydes.

Reactions of carbonyl groups

Monosaccharides like glucose, mannose and fructose form osazones by reacting with two molecules of phenyl hydrazine. The reaction takes place between the first two carbon atoms of the sugars. Osazones are insoluble compounds and are used for the identification of sugars. The osazone reaction is useful in identifying aldose sugars differing in configuration at the α carbon. The upper equation shows the general form of the osazone reaction, which effects an alpha-carbon oxidation with formation of a bis-phenylhydrazone, known as an osazone. Application of the osazone reaction to D-glucose and D-mannose shows that they differ in configuration only at C-2.

1.

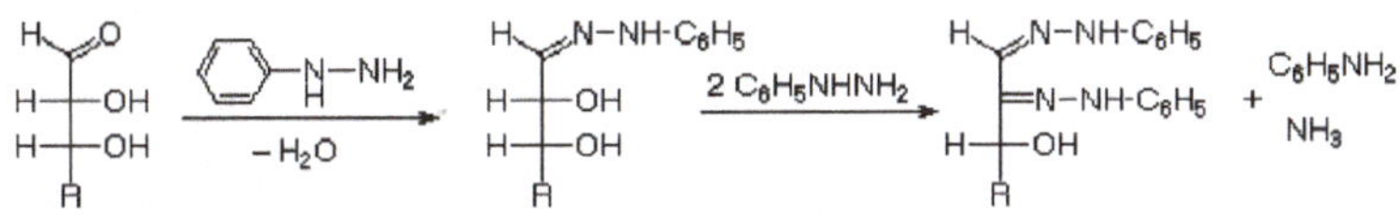

2.

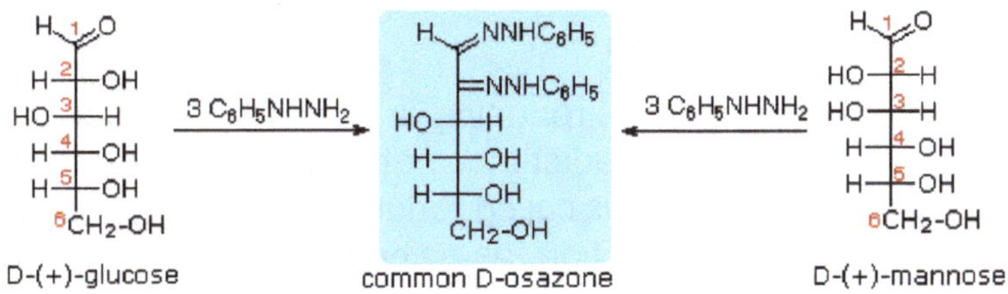

Chain Shortening and Lengthening

These two procedures permit an aldose of a given size to be converted to homologous smaller and larger aldoses. The importance of these relationships may be seen in the array of aldose structures. The degradation of the pentose arabinose gives the tetrose erythrose. Working in the opposite direction, arabinose can be converted to a mixture of glucose and mannose.

1.

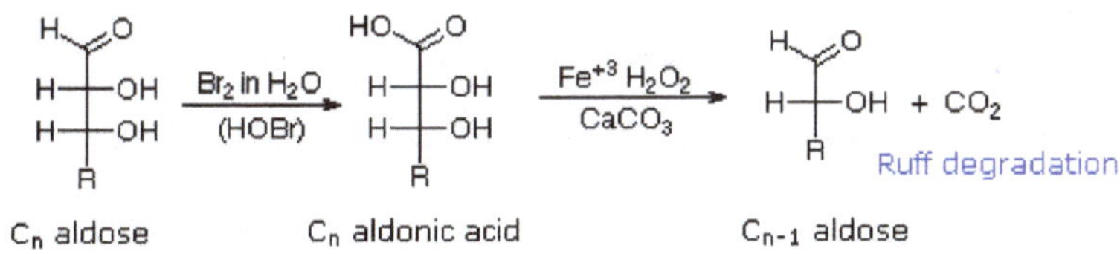

2.

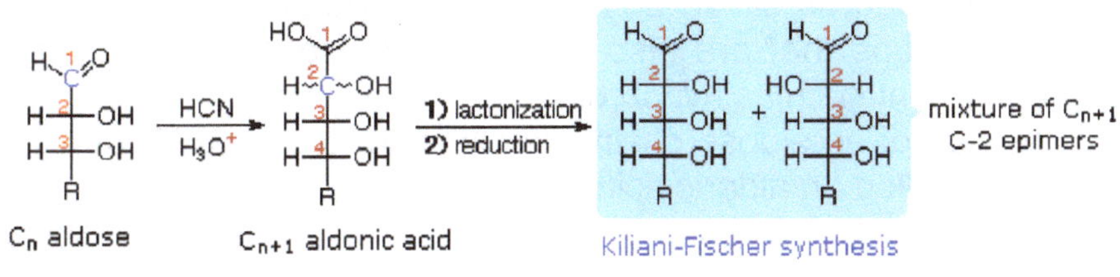

The carbonyl group reacts with HCN to form cyanhydrins. This reaction is used to prepare sugars of higher homologous series. Sugars on reduction can be converted to their respective alcohols.

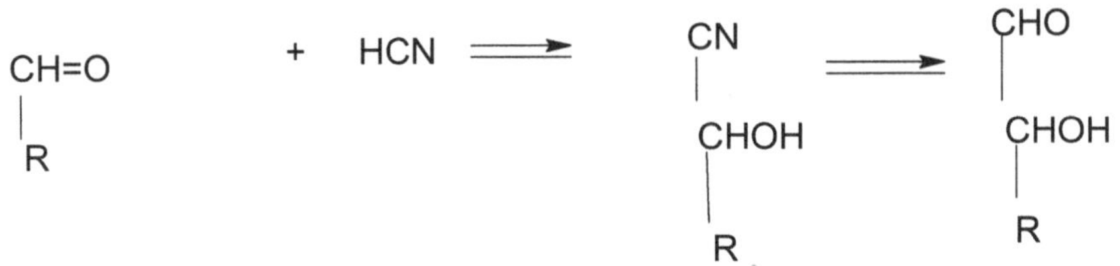

Formation of higher homologue of sugar

Oxidation and reduction

Sugars can be classified as reducing or non-reducing based on their reactivity with Tollens', Benedict's or Fehling's reagents. If a sugar is oxidized by these reagents it is called reducing, since the oxidant ($Ag^{(+)}$ or $Cu^{(+2)}$) is reduced in the reaction, as evidenced by formation of a silver mirror or precipitation of cuprous oxide. The Tollens' test is commonly used to detect aldehyde functions; and because of the facile interconversion of ketoses and aldoses under the basic conditions of this test, ketoses such as fructose also react and are classified as reducing sugars.

Reactions with different oxidizing agents are used to detect and estimate the reducing sugars. The frequently used reagents are Tollens, Benedicts and Fehlings' solutions. Tollens test is based on the reduction of Ag ions and the other two on the reduction of Cu ions.

1. –CHO + Ag $(NH_3)_2$ (Tollens Reagent) $\rightarrow$ -$COOH^-$ + Ag

2. –CHO + Cu (citrate—Benedicts blue) $\rightarrow$ -$COOH^-$ + CH_2O (brick red)

3. –CHO + Cu (tartrate-Fehlings-blue) $\rightarrow$ -$COOH^-$ + CH_2O (brick red)

When the aldehyde function of an aldose is oxidized to a carboxylic acid the product is called an aldonic acid. Because of the two end hydroxyl groups that are also present in these compounds, a mild oxidizing agent such as hypobromite must be used for this conversion. If both ends of an aldose chain are oxidized to carboxylic groups the product is called an aldaric acid.

Glucose on reduction gets converted to sorbitol and on oxidation changes to glucuronic acid (mono carboxylic) followed by glucosaccharic acid (dicarboxylic). Oxidizing agents like nitric acid oxidize aldehyde group as well as alcohols to sugar acids called saccharic acids.

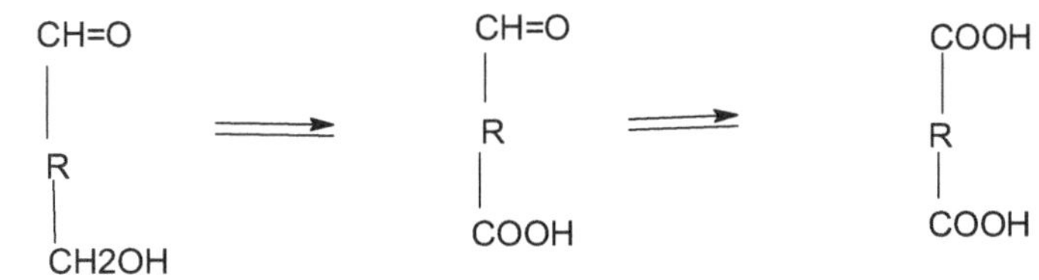

GLUCOSE D-GLUCURONIC ACID D-GLUCOSACCHARIC ACID

Sodium borohydride reduction of an aldose makes the ends of the resulting alditol chain identical, $HOCH_2(CHOH)_nCH_2OH$, thereby accomplishing the same configurational change produced by oxidation to an aldaric acid. Thus, alditol and galactitol from reduction of allose and galactose are achiral, and altrose and talose are reduced to the same chiral alditol. A summary of these redox reactions, and derivative names are given in the following table.

Table 2. Oxidation and reduction reactions of sugars

	Derivatives of $HOCH_2(CHOH)_nCHO$	
HOBr Oxidation	⟶	$HOCH_2(CHOH)_nCO_2H$ an Aldonic Acid
HNO_3 Oxidation	⟶	$H_2OC(CHOH)_nCO_2H$ an Aldaric Acid
$NaBH_4$ Reduction	⟶	$HOCH_2(CHOH)_nCH_2OH$ an Alditol

Reactions of alcoholic groups

1. **Formation of glycosides:** Sugars react with alcohols to form acetals.

$$RCH\text{-}OH + B\text{-}OH \longrightarrow RCH\text{-}OB + H_2O$$

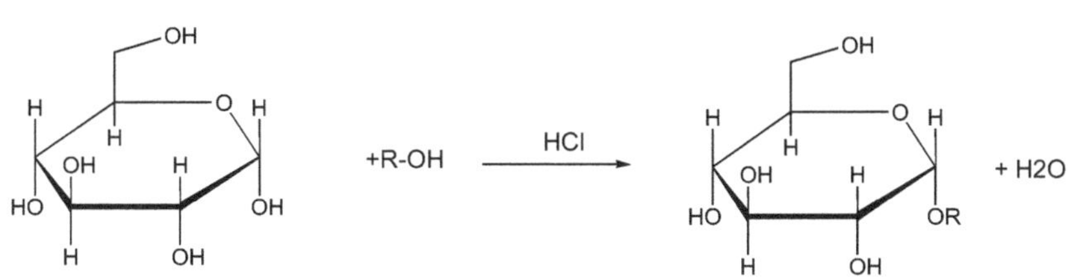

ALPHA D- GLUCOPYRANOSE GLYCOSIDE - AN ACETAL

2. **Esters:** Alcohol group reacts with acids to form esters. Ex. Glucose-1 phosphate, glucose-6 phosphate, fructose phosphate etc.
3. **Dehydration:** H_2SO_4 / HCl treatment produces furfurol

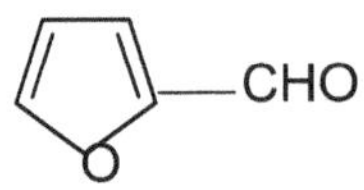

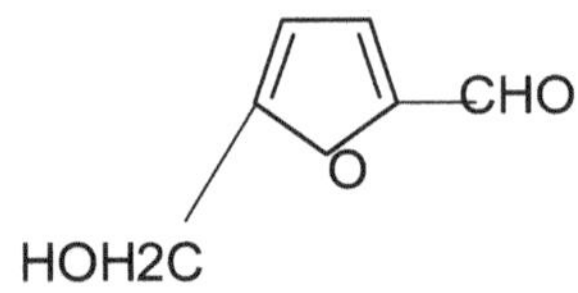

FURFUROL HYDROXYETHYLFURFUROL

4. **Amino-sugars:** Amino- sugars are those in which a hydroxyl group in a sugar is replaced by an amino group. The common amino sugars are Glucosamine (2-amino –D-glucose) and galactosamine. In some cases the amino group is acetylated. For ex. N- Acetylgalactosamine.

Uses of carbohydrates

Monosaccharides, oligosaccharides and polysaccharides are used in a number of industries as listed below:

Monosaccharides

1. Glucose and fructose are used as energy source
2. Liquid glucose is widely used in the confectionary, bakery, and jam preparation, canning and leather industries.
3. Glucose can be fermented to the biofuel ethanol.
4. Liquid dextrose is used in fermentation industries, for the manufacture of dextrose monohydrates, fructose and sorbitol syrups.
5. Sorbitol syrup is widely used in tooth paste, pharmaceuticals, cosmetics and tobacco industries.
6. Fructose is used as sweetener in beverages, sport drinks and also used as a flavoring agent.
7. Fructose is used in cosmetic and pharmaceutical industries

Oligosaccharides

1. Sucrose is used in confectionery industry and in desserts.
2. Sucrose is used in preservation of foods.
3. Sucrose is used in cosmetic and pharmaceutical industries.
4. Maltose is used in baby food industry.

Polysaccharides

1. **Food industry:** Starch plays a leading role in determining the texture of many foods. Texture is of vital concern to both the consumers and the manufacturers. Hydrolysis of starch, usually by enzymatic reactions, produces a syrupy liquid consisting largely of glucose. When corn starch is the feedstock, this product is known as corn syrup. It is widely used to soften texture, add volume, prohibit crystallization and enhance the flavor of foods. Starch finds numerous uses in the baking industry for the production of cakes, cookies, in ice-cream preparations etc.
2. **Paper industry:** In paper industry, a large quantity of starch is consumed as a surface-sizing agent, as a binder, as a paper coating agent etc. Starch is used in the manufacture of various adhesives or glues for book-binding, wall paper adhesives, gummed paper, envelop adhesives, school glues and bottle labeling.
3. **Textile industry:** In textile industry, starch is used in sizing to strengthen the warp yarn, in finishing and changing the appearance of fabric after it is bleached, dyed or printed. Starch is used as a component in finishing agent to glaze and polish sizing thread. Clothing starch or laundry starch is a liquid that is prepared by mixing a vegetable starch in water.
4. **Pharmaceutical industry:** Starch is used as an excipient, a binder in medications to aid the formation of tablets
5. **Printing industry:** In the printing industry, food grade starch is used in the manufacture of anti-set-off spray powder used to separate printed sheets of paper to avoid wet ink being set off.
6. **Plastic industry:** Starch is used to produce various bio-plastics, synthetic polymers that are biodegradable.

Uses of cellulose

Cotton, the most useful natural fiber, is nearly pure cellulose. The manufacture of textiles from cotton involves physical manipulation of the raw material by carding, combing and spinning selected fibers. For fabrics the best cotton has long fibers, and short fibers or cotton dust are removed. Crude cellulose from wood pulp is widely used for making paper. Solutions of cellulose and their derivatives can be spun into fibers.

Cellulose Nitrate, called guncotton, is explosively flammable and is a component of smokeless powder. Partially nitrated cellulose is called pyroxylin finds uses in photographic film and lacquers. Cellulose Acetate, —$[C_6H_7O(OAc)_3]_n$—, is less flammable than pyroxylin, and has replaced it in most applications.

An acetone solution of cellulose acetate may be forced through a spinneret to generate filaments, called acetate rayon, that can be woven into fabrics. Viscose Rayon, is prepared by formation of an alkali soluble xanthate derivative that can be spun into a fiber that reforms the cellulose polymer by acid quenching. The product fiber is called viscose rayon.

Chapter – 3

Lipids

Lipids are non polar substances of biological origin that are sparingly soluble in water, but soluble in organic solvents like chloroform, hexane, benzene etc. The word lipid comes from the Greek word *lipos* meaning lard or fat. The lipids present in biological systems are either hydrophobic (non polar) or amphipathic (both polar and non polar). Lipids like fats, oils or glycerophospholipids can be hydrolyzed to get fatty acids and other products. Hydrophobic nature is useful in making lipids as membranes which act as effective barriers to more polar molecules. Steroids are also considered lipids, but cannot be hydrolysed and are components of non-saponifiables in fats. Waxes, which are esters of fatty acids with long chain alcohols are also considered as lipids.

Fats or oils are organic esters of fatty acids and glycerol. Hence they are called triglycerides or triacyl glycerols. Most plant lipids are acyl glycerols. Fats are solid at room temperature, but when they are in the liquid state, they are called oils. Lipids and oils are important constituents of food and account for their higher calorific value. As molecules that are largely hydrocarbon in nature, lipids represent highly reduced forms of carbon and upon oxidation in metabolism, yield large amounts of energy. Fats are the richest sources of energy on a weight basis. Their hydrophobic nature allows them to aggregate in highly anhydrous forms. Lipids are thus the molecules of choice for metabolic energy storage.

Fats and oils provide on oxidation 9kcal/g (38kJ/g) energy as against 4kcal/g (17kJ/g) of either sugars or proteins. They are sources of fat soluble vitamins and essential fatty acids to human beings. When the supply of energy is in excess, it is stored in the form of lipid in the adipose tissue of the human being. Body fat provides insulation. As a storage fuel, fats have a special advantage. They can be stored in an anhydrous form as they are hydrophobic while glycogen is hydrated with lesser energy value. Inspite of fat being misunderstood as something of larger size, fat as a source of energy occupies lesser space than either sugars or proteins. A gram of lipid can store six times more energy than a gram of glycogen. Lipids are important to the structure, composition and permeability of membranes and cell walls.

Triglycerides are widely distributed in the plant kingdom. In higher plants they are present in vegetative and reproductive tissues. The vegetative tissues contain small quantities while the reproductive tissues , namely fruits and seeds contain large amounts. The lipids found in biological systems are either hydrophobic (non-polar) or amphipathic (polar and non-polar). The hydrophobic nature of lipid molecules allows membranes to act as effective barriers to more polar molecules. Membranes act as boundaries around cells and cell compartments. They are semi-permeable and are involved in signaling processes.

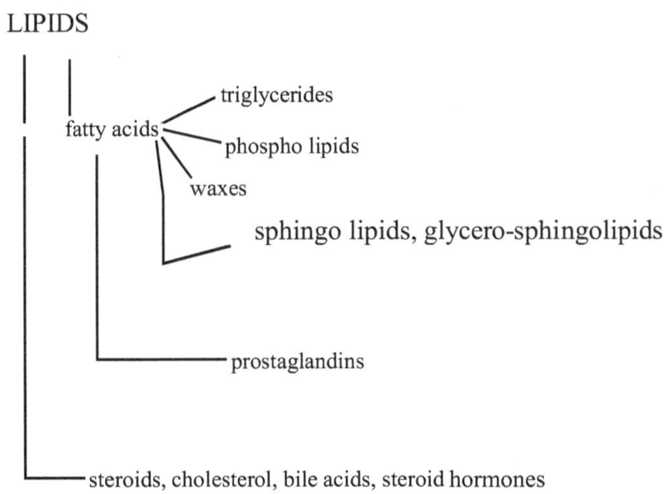

Classification of lipids

Lipids can be classified based on various criteria. A widely accepted criterion for lipid classification is their constituent units. Accordingly, the lipids are classified into two major groups, namely, simple and complex lipids.

1. Simple or non-saponifiable lipids: Lipids belonging to this class cannot be hydrolyzed (saponified) to form simpler units. They do not have any backbone too. Terpenes, steroids and prostaglandins are the main non-saponifiable type of lipids.

Terpenes: Terpenes are made up of isoprene units. Terpenes are the most widespread class of biochemical substances in plants.

Steroids: Steroids are predominant mostly in eukaryotes. They are derivatives of a ring referred to as cyclopentano perhydro phenanthrene made up of three six and one five membered fused rings. Substitution and opening up of the five membered ring forms different steroids. The most important steroid is cholesterol because of the direct relationship between the incidence of cardiovascular problems and the level of serum cholesterol in human beings. Plants however, contain very little cholesterol, while it is the metabolic precursor of many sex harmones in humans. However, many phytosterols like sitosterols, stigmasterol, brassinosterols etc. are present in plants.

***Prostaglandins*:** Prostaglandins are synthesized from the unsaturated fatty acid, namely, arachidonic acid (20:4, ^{5,8,11,14} all *cis*). Prostaglandins are of significance in mammals in lowering blood pressure, controlling inflammatory response in skin and joints, and induction of labor etc. These are short lived organic molecules which act near their place of synthesis.

2. Complex or saponifiable lipids: Lipids belonging to this class can be hydrolyzed (saponified) into two or more constituent units. They are further sub-classified depending upon their backbone constituent.

Table 1. Different lipid types

S. No.	Sub-Class	Backbone
1.	Acyl glycerols	Glycerol
2.	Phosphoglycerides	Glycerol-3-phosphate
3.	Sphingolipids	Sphingosine (18 C alcohol)
4.	Waxes	Long chain monohydroxy alcohols

Acyl glycerols: Acyl glycerols are made up of the alcohol glycerol and fatty acids. Glycerol is a trihydroxy alcohol. A triacyl glycerol is formed by esterification, in which the hydroxyl groups of glycerol form ester bonds with the carboxyl groups of fatty acids. For example, glycerol reacts with three molecules of palmitic acid to form tripalmitin or three molecules of oleic acid to form triolein etc. Glycerol can esterify with three molecules of a single fatty acid or three different fatty acid molecules. Triacyl glycerols are the major forms of energy storage for animals. During times of crisis and starvation stored fat is the only source of energy. Fats stored in plant seeds serve as sources of energy during germination.

Fatty acids: Fatty acids are the carboxylic acids with a hydrocarbon chain. They are the most important constituents of lipids. They contain normally 12 to 26 Carbon (C) atoms; however, 16 and 18 C fatty acids are the most common. Most fatty acids present in lipids contain an even number (12, 14, 16, 18, 20, etc.) of carbon atoms. They may be linear or branched or may have cyclic groups. They may have additional functional groups like OH, C=O (keto) etc. They may be saturated fatty acids (SFAs) with no double bond or unsaturated fatty acids (USFAs) containing up to six double bonds. SFAs up to 8 C atoms are liquids at room temperature, while those with more than 8 C atoms are solids. The melting point of fatty acids therefore increases with increase in chain length.

SFAs: They can be represented by the general formula $CH_3(CH_2)_n$ COOH. The simplest (n=0) is acetic acid. Acids having up to 10 carbon atoms are considered simple organic acids. Those having 10 and above carbon atoms are called fatty acids. They occur mostly in the combined form and rarely in the free form. The most widely occurring fatty acids have 12 to 18 C atoms. Those above 20 C atoms occur in lesser quantities.

USFAs: They may have 1 to 6 double bonds in their structure. They have a methylene (CH_2) group between the double bonds. In the naturally occurring USFAs, the double bonds are of *cis type*. The USFAs with two or more double bond are referred to as polyunsaturated fatty acids (PUFAs). Among the USFAs, oleic, an 18 C fatty acid with one double bond (18:1) and linoleic acid, an 18 C fatty acid with two double bonds(18:2) are the most abundant followed by linolenic acid, the 18C fatty acid with 3 double bonds (18:3). In the symbols 18:1, 18:2 and 18:3 the number 18 indicates the number of carbons and the number 1,2 and 3 indicate the number of double bonds. If 0 is used after the carbon number, it indicates zero unsaturation or a saturated molecule (SFA).

The USFAs are more condensed in length than the SFAs of equal chain length. The double bonds give the fatty acid molecules a sharp bend. The double bonds thus prevent tight packing within the membranes.

This behavior of USFAs does have profound biological significance. Further, the introduction of double bonds in fatty acids decreases the melting point (M.P.) of the fatty acids. Thus, 18 C saturated fatty acid – Stearic acid has a melting point of 69 °C, 18 C unsaturated fatty acid – Oleic acid (one double bond) has 13 ⁰C and 18C unsaturated fatty acid – Linoleic acid (two double bonds) has a melting point of – 17 ⁰C. The USFAs enter into oxidative and polymerization reactions. They cause rancidity and even promote tumors.

Table 2. Chemical structure and names of some common fatty acids

Common name	Chemical structure, and symbol	Systematic name
Lauric acid,	$CH_3(CH_2)_{10}COOH$ (12:0)	Dodecanoic acid
Myristic acid,	$CH_3(CH2)_{12}COOH$ (14:0)	Tetradecanoic acid
Palmitic acid,	$CH_3(CH2)_{14}COOH$ (16:0)	Hexadecanoic acid
Stearic acid,	$CH_3(CH2)_{16}COOH$ (18:0)	Octadecanoic acid
Arachidic acid,	$CH_3(CH2)_{18}COOH$ (20:0)	Eicosanoic acid
Behenic acid,	$CH_3(CH2)_{20}COOH$ (22:0)	Docosanoic acid
Lignoceric acid,	$CH_3(CH2)_{22}COOH$ (24:0)	Tetracosanoic acid
Palmitoleic acid,	$CH_3(CH2)_5CH{=}CH(CH_2)^7$ COOH (16:1) Δ9 *cis* 9 –	Hexadecenoic acid
Oleic acid,	$CH_3(CH^2)_7CH{=}CH(CH_2)^7COOH$ (18:1) Δ9 *cis* 9 –	Octadecenoic acid
Linoleic acid,	$CH_3(CH^2)_4$ $(CH{=}CH\ CH_2)^2$ $(CH_2)^3$ COOH (18:2)Δ9,12 both *cis*	Octadecadienoic acid
α- Linolenic acid,	CH_3 CH_2 $(CH{=}CH\ CH_2)^3$ $(CH_2)_6$ COOH (18:3)Δ9,12,15 all *cis*	Octadecatrienoic acid
Ý-Linolenic acid	$CH_3(CH2)_4$ $(CH{=}CH\ CH_2)_3$ (CH2)3COOH (18:3)Δ6,9,12 all *cis*	Octadecatrienoic acid
Arachidonic acid	$CH_3(CH_2)_4(CH{=}CH\ CH_2)$ $(CH_2)_2$ -COOH (20:4) Δ 5,8,11,14	all *cis eicosa tetra*enoic acid
Ricinoleic acid,	$CH_3(CH_2)_5CHOHCH_2CH{=}CH$ $(CH_2)_7COOH$ (18:1 Δ9 *cis* 12-OH)	12- Hydroxyoctadecenoic acid

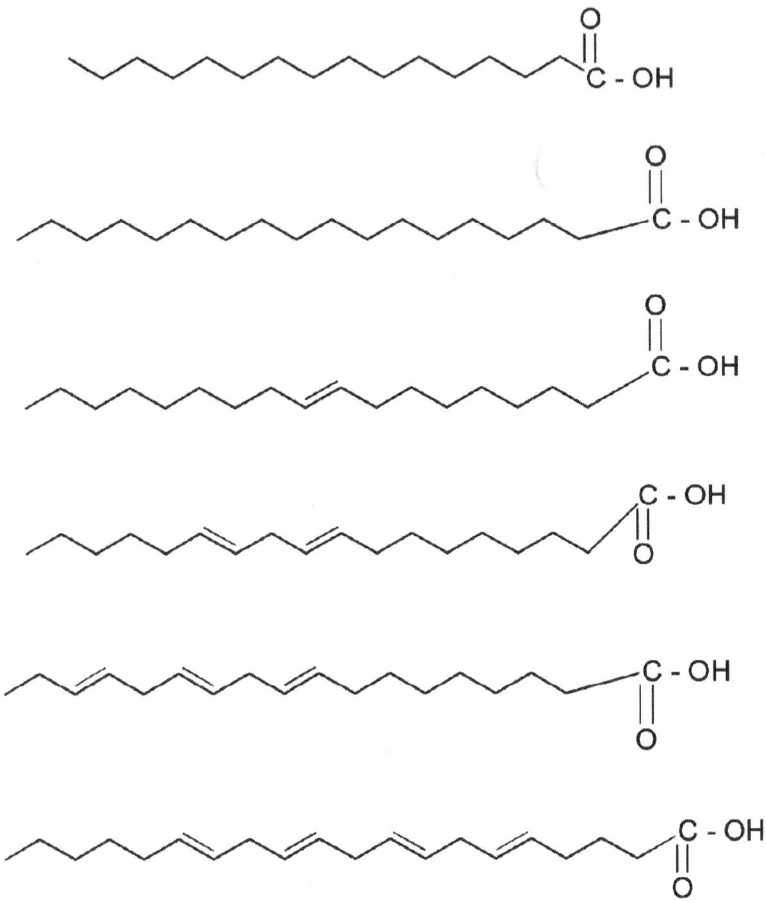

16:0

18:0

18:1

18:2

18:3

20:4

Structure of some fatty acid

Essential fatty acids (EFAs): The USFAs viz. linoleic, (α and Ý) linolenic and arachidonic (20:4) acids are not synthesized by mammals and are therefore important dietary requirements for good health and growth. Hence they are called Essential Fatty Acids. They are precursors of prostaglandins and related hormones in the human body. They have to be derived from plant sources. Arachidonic acid is not present in plants and can be synthesized from linoleic acid. EFA deficiency in humans results in abnormal skin conditions such as scaliness and dermatitis. It also causes increased water loss and increased susceptibility to infection. The recommended daily intake of EFA for an adult human being is 3% of total energy or 2-10g per day of linoleic acid.

Acyl glycerols

In acyl glycerols, one, two or three OH groups of glycerol are esterified by fatty acids (RCOOH) to form mono -, di -, and triacyl glycerols respectively. Mono-, di- and triacyl glycerols (MAG, DAG and TAG), are commonly referred to as mono-, di-and triglycerides, respectively. If all the three –OH groups are esterified with 'one or single type' fatty acid, the triglycerides are called simple triglycerides. If two or three different fatty acids are involved, they are called mixed triglycerides. For example, if all the three OH- groups are esterified with palmitic acid it is called tri-palmitin, a simple triglyceride. If the three OH-groups of glycerol are esterified with palmitic, stearic and oleic acids, it is called palmito-oleo-stearin, a mixed triglyceride. Natural glycerides are mixtures of mixed triglycerides. They occur as liquids or solids depending upon their constituent fatty acids. Triacyl glycerols include fats (solids at room temperature) and oils (liqiuds at room temperature). MAG and DAG are components of many biological membranes.

Glycerol+ Fatty acids = Triacyl glycerols or Triglycerides(oils or fats)

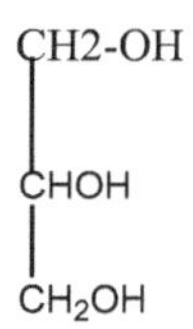

GLYCEROL

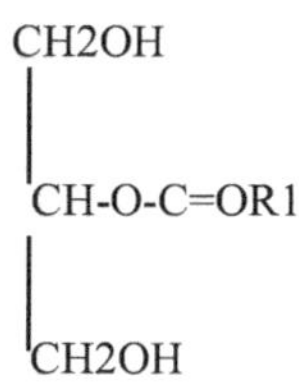

2-MONOGLYCERIDE

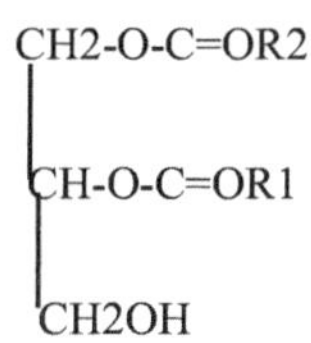

1,2 DIGLYCERIDE

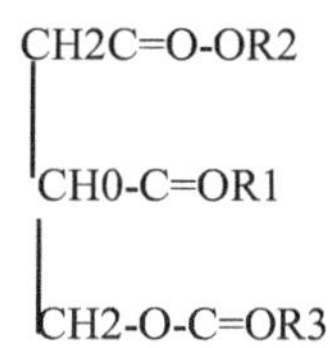

1,2,3 TRIGLYCERIDE

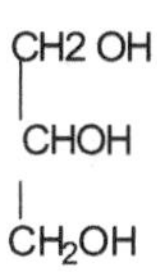

GLYCEROL

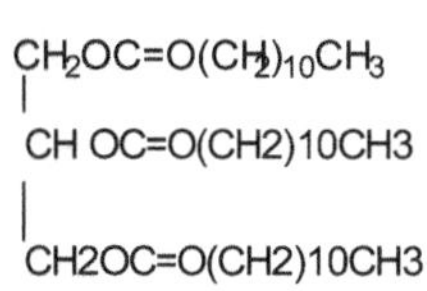

GLYCERYL LAURATE
A SIMPLE TRIGLYCERIDE

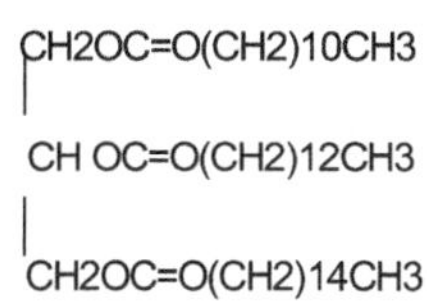

GLYCERYL PALMITO MYRISTO
LAURATE–A MIXED TRIGLYCERIDE

Phosphoglycerides (phospholipids): They are widespread in nature. The parent compound is phosphatidic acid – glycerol 3- phosphate in which esterification of C1 and C2 -OH groups are by fatty acids and that of C3 OH is by phosphoric acid. The phosphate in turn remains joined (esterified) to a range of substances containing OH groups forming various phosphoglycerides. Phosphoglycerides are amphipathic molecules with a non polar aliphatic tail and a polar phosphorylated head. In a bilayer arrangement these form the major components of biological membranes within which proteins remain embedded. They are essential components of cell membranes and are found in small concentrations in other parts.

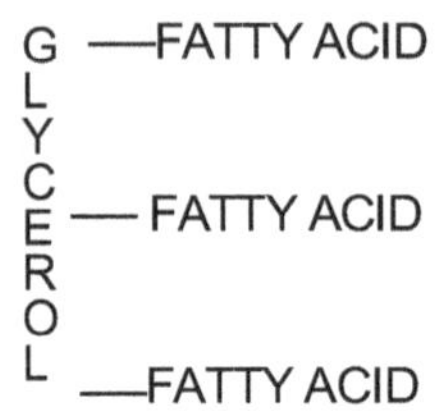

TRIACYL GLYCEROL

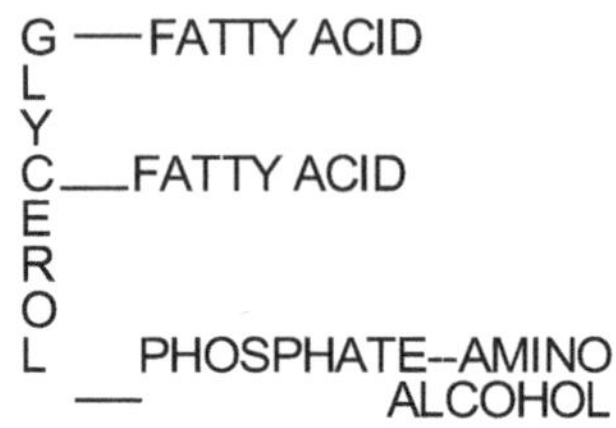

GLYCERO PHOSPHO LIPID

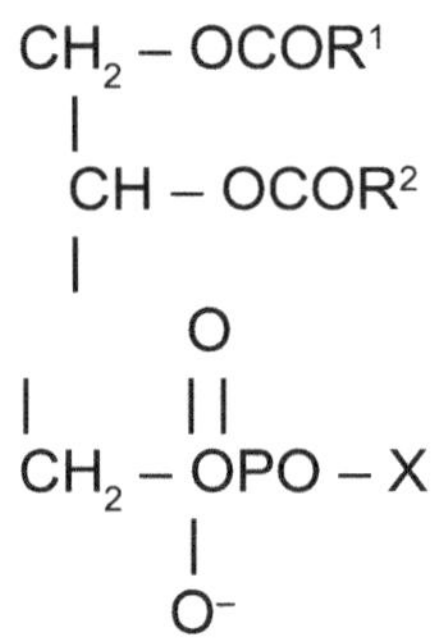

If X = H the phospholipid is phosphatidic acid

If X = choline, HOCH2CH2N+(CH3)3, it is phohatidyl choline (lecithin)

If X = ethanolamine, HOCH2CH2NH2, it is phohatidyl ethanolamine (cephaline)

If X = serine, HOCH2CHNH2COOH, it is phohatidyl serine

If X = glycerol, CH2OHCHOHCH2OH, Phohatidyl glycerol

If X = myolinositol, it is called phosphtidyl inositol (cadiolipin)

Choline, serine and ethanolamine are the three amino alcohols present in glycerol phospholipids. They are ionized at the body physiological temperature of p^H 7.4. Lecithins and cephalins are present in brain and nerve tissue as well as egg yolk, wheat germ and yeast. Lecithins contain choline and cephalins contain ethanolamimine and sometimes serine.

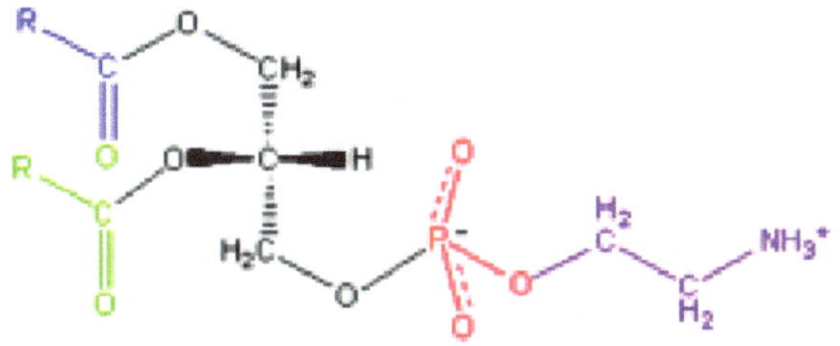

Phosphatidyl ethanolamine

Phospholipids are present in plant and animal cells, especially in the photosynthetic tissue of plants. They are of importance in insulating the nerve impulse (like the insulator around an electric wire) from the surrounding structures and in channeling out enzymes. They are also involved in transport of lipids in the blood stream. Cephalins and lecithins are found in brain tissue and nerves. Eggs and some oilseeds are rich in lecithins. Cephalins are involved in blood clotting. Snake venom hydrolyses one of the fatty acid ester unit of lecithin. The resulting product causes breakdown of blood cells.

Phospholipids are often called polar lipids. They undergo hydrolysis when heated with acids to yield fatty acids, glycerol, phosphoric acid and the amino alcohol. Phospholipids are the major components of membrane lipids. The most abundant fatty acids in phospholipids are 16:0, 18:0 and 18:1. Phospholipids are usually complexed and present as lipoproteins in nature. They are formed in the plastids and endoplasmic reticulum.

Sphingolipids : Sphingolipids are also important components of membranes. They however, do not contain glycerol. Instead, they contain an 18 C amino alcohol called sphingosine or its derivatives. Sphingolipids are important components in nerve cell membranes where they protect and insulate the nerve fibres. Glycosphingolipids are sphingolipids that contain galactose or glucose. They are components of nerve membranes and muscle.

Waxes : Waxes are water insoluble substances present on the surface of plant and animal bodies. Waxes occur in the cuticles of plants. They include a variety of compounds such as hydrocarbons, alcohols, ketones, and carboxylic acids and most importantly the esters of long the chain monohydroxy alcohols (24 to 28 C) with fatty acids (20 to 24 C). Fatty acids, fatty alcohols and fatty aldehydes accompany wax esters. The cuticle layer of tropical plant leaves is heavier due to the presence of saturated compounds and the waxes are high melting.

Waxes are used as carriers in cosmetics and a variety of coloring materials. Waxes contain many unusual fatty acids. Waxes are water repelling and prevent loss of water and damage from insects in plants. Shining appearance of fruits, leaves, and flower petals is often due to the presence of waxes. Lotus leaves contain high amounts of waxes. Waxes on the skin, fur and feathers of animals and birds provide a water proof coating. Examples of natural waxes are bees wax, carnauba wax, jojoba wax, ear wax etc.

Table 3. Some important waxes

Type of wax	Formula	Occurance	Uses
Bees wax	CH3(CH2)18COO(CH2)19CH3	Honey bee	Candles,shoe polish, wax paper
Carnauba wax	CH3(CH2)24COO(CH2)29CH3	Palm tree	Waxes for shoes, furniture,cars, floors
Jojoba wax	CH3(CH2)18COO(CH2)19CH3	Jojoba	Candles, soaps, cosmetics

Prostaglandins : These are hormone- like substances present in most cells of the body. Prostaglandins are also called as eicosanoids as they are derivatives of C20 fatty acid, namely arachidic acid (20:0). They have a lot of physiological effects. Some of them increase blood pressure(B.P.). while some decrease B.P. They stimulate contraction and relaxation of smooth muscles such as uterus. Hence, they are used in child birth. In larger doses they can cause abortion. They are also used in treatment of gastric ulcers and bronchial asthma. They also inhibit platelet aggregation. When a tissue is injured, prostaglandins are produced with concomitant pain and inflammation. Normally they are not in the mammalian tissues, but are mostly produced in response to stimuli.

Membrane lipids: Membranes which separate cells contain lipids and small quantities of carbohydrates. They act as permeable membranes. Membranes contain glycerophospholipids, shingolipids and sterols. Glycerophospholipids have glycerol backbone with two fatty acids and a phosphorylated group attached to it. The latter group may contain phosphatidate, phosphatidylcholine, phosphatidylethanolamine, phosphatidylglycerol,phosphatidylinositol,phospatidylserine or diphophatidylglycerol. The prominent phospholipid in the membranes is phosphatidyl glycerol. Membrane lipids are amphiphilic since they contain both hydrophilc and hydrophobic groups. Lipids orient in such a way that the hydrophobic areas orient inside while the hydrophilic groups are on the surfaces. Since there are no covalent bonds between the bilayers, the membranes have fluidity.

The glycerophospholipids are the main structural components of biological membranes, such as the cellular plasma membrane and the intracellular membranes of organelles. However, other non-glyceride lipid components such as sphingomyelin and sterols are also found in biological membranes. In plants and algae, the galactosyldiacylglycerols, and sulfoquinovosyldiacylglycerol, which lack a phosphate group, are important components of membranes of chloroplasts and related organelles and are the most abundant lipids in photosynthetic tissues, including those of higher plants, algae and certain bacteria. Chloroplasts contain protein-glycolipid structures.

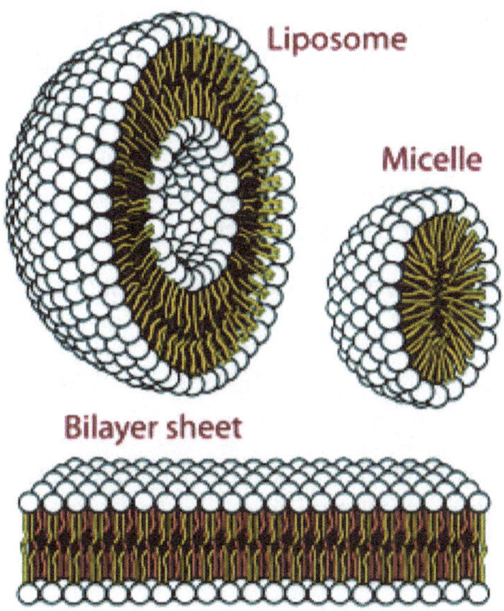

A spherical liposome, a micelle, and a lipid bilayer

A biological membrane is a form of lipid bilayer. The formation of lipid bilayers is an energetically preferred process when the glycerophospholipids are in an aqueous environment. This is known as the hydrophobic effect. In an aqueous system, the polar heads of lipids align towards the polar, aqueous environment, while the hydrophobic tails minimize their contact with water and tend to cluster together, forming a vesicle. Micelles and bilayers form in the polar medium by a process known as the hydrophobic effect.

Functions of lipids

Lipids encompass a very large and diverse group of biological substances and so they perform wide ranging functions too. Some major functions of lipids are as follows:

1. Lipids act as reservoirs of energy in biological systems. Oils and fats should form 20-30% of food. When consumed as food they are more satisfying and cause delay in hunger. Being more reduced than carbohydrates, lipids can store more energy. The most important storage form of lipids is the triacylglycerols in the oil bodies in plant seeds and adipose tissues in animals. They release 9kcal/g of energy as against 4kcal/g by either proteins or sugars.
2. Lipids act as the major components of biological membranes. The most Important class of lipids in this regard is the amphipathic phospholipids with a small hydrophilic head and a long hydrophobic tail arranged in a bilayer form.
3. Some lipids act as members of electron transport system in inner mitochondrial membrane viz. ubiquinone and also phosphorylation systems in thylakoid membrane.
4. Lipids act as carriers of sugars viz. dolichol in the biosynthesis of glycoproteins.
5. Lipids are used for the biosynthesis of certain hormones in animals and plants.
6. Lipids in the form of bile acids (e.g. cholic acid) help in the digestion and absorption of other lipids.
7. Triacyl glycerols act as heat insulating materials.
8. Members of several different lipid categories have been identified as signaling molecules and cellular messengers. For example, sphingosine-1-phosphate, a sphingolipid, diacylglycerol (DAG) and the phosphatidylinositol phosphates, the prostaglandins, which are involved in inflammation and immunity and the steroid hormones such as estrogen, testosterone and cortisol, which modulate a host of functions such as reproduction, metabolism and blood pressure.
9. Llipids are sources and carriers of vitamins like A,D,E and K.

Properties of fatty acids

The lower fatty acids are liquid at room temperatures. Some of them are soluble in water and are steam volatile. The higher fatty acids are solid and are insoluble in water.

1. *Fatty acids form esters with alcohols* : The esters with glycerol are what we call glycerides or oils. Esters with long chain monohydric alcohols are the waxes.

   $RCOOH + C_2H_5OH \rightarrow RCOO\ C_2H_5 + H_2O$

2. Formation of soaps with alkalies:

   $RCOOH + NaOH \rightarrow RCOONa + H_2O$

   Alkali salts of fatty acids are soaps: The sodium and potassium salts of fatty acids are the ones which are used daily. Sodium soaps are hard while potassium soaps are soft. Zinc stearate, which is non-irritant is used in dusting powders.

3. Hydrogenation : USFAs combine with hydrogen to form saturated fatty acids.

   $RCH{=}CH\text{———}COOH + H_2 \rightarrow RCH_2CH_2\text{———}COOH$

4. Halogenation : The double bond can be saturated with chlorine/ bromine/iodine.

   $RCH{=}CH\text{———}COOH + I_2 \rightarrow RCHI\text{-}CHI\text{———}COOH$

5. Isomerism at the double bond: In addition to the normal isomers due to the asymmetric carbon, *cis* and *trans* isomers are possible at the double bond. In the *cis* position both the groups at the double bond are on one side while in the *trans* position the groups are on the opposite side.

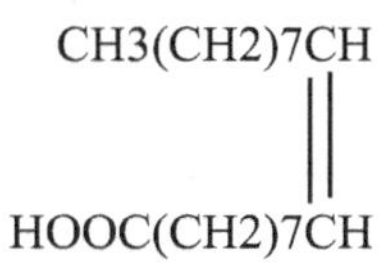

CIS- OLEIC ACID

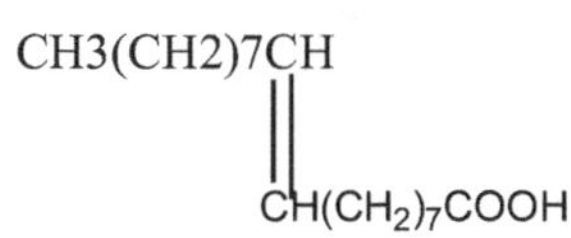

TRANS- ELAIDIC ACID

Properties of fats/ oils (acylglycerols)

Fats/oils are colorless, odorless and tasteless compounds. They are insoluble in water and soluble in organic solvents. They have low specific gravity (0.9-0.95) and hence float on water.

Hydrolysis: Superheated steam, acids and alkalies hydrolyze fats to fatty acids and glycerol.

Additive reactions: Unsarturated fats participate in additive reactions with hydrogen and halogens similar to the USFAs. Liquid oils are hydrogenated to solid fats (*Vanaspati-Dalda*), which have higher storable/ shelf life.

Oxidation: Fats rich in unsaturated fatty acids undergo oxidation. Unsaturated oils like linseed oil are useful as drying oils. They undergo oxidation and polymerization to change to thin films (paints & varnishes).

Rancidity of oils and fats

Unsaturated fats get hydrolysed or oxidized in the presence of oxygen, water and some enzymes (lipases). Oils develop foul or bad smell which is what we call rancidity. Higher unsaturation causes quicker rancidity. Vitamin E and other antioxidants present in oils prevent oxidation and thus the rancidity. Tocopherols (vitamin E) are hence called auto-oxidative stabilizers.

Rancidity is of two types. They are a) hydrolytic and b) oxidative rancidities. Hydrolytic rancidity is due to the hydrolysis and release of free fatty acids. Oxidative rancidity, however, is more prevalent. It is promoted by heat, light, ionizing radiation, catalysts and enzymes. Fats containing 18:2 and 18:3 are more prone to oxidation and rancidity than fats having 18:1.

Rancid oils are rich in free fatty acids (FFA). The smoke point of oils gets reduced when the FFA is higher. For Ex. cottonseed oil with 1% FFA has a smoke point of 160^0 C as against 232^0 C for the neutral oil. Foods fried in higher FFA oils show cracking on surfaces, more browning and more absorption of oil.

Hydroperoxides, hydroxy acids, aldehydes and ketones are formed during oxidation of fats. Free radicals produced during rancidity have a role in cancer. Butter on rancidity produces foul smelling butyric, caprylic and caproic acids. Rancid foods usually get rejected due to their foul smell.

Rancidity in oils can be prevented by adding some antioxidants like butylated hydroxy toluene(BHT) or butylated hydroxy anisole (BHA), vitamin C or vitamin E.

Analytical properties of lipids

Saponification: The term 'Saponification' refers to the alkaline hydrolysis of saponifiable lipids , most importantly the fats and oils. In the case of fats and oils, the products are glycerol and the salts of the respective fatty acids. Thus, R COONa and R COOK (sodium palmitate, sodium stearate, potassium laurate etc.) are the akali salts of the fatty acids. The reaction is utilized in the making of soaps. Sodium soaps are somewhat hard soaps while potassium soaps are soft.

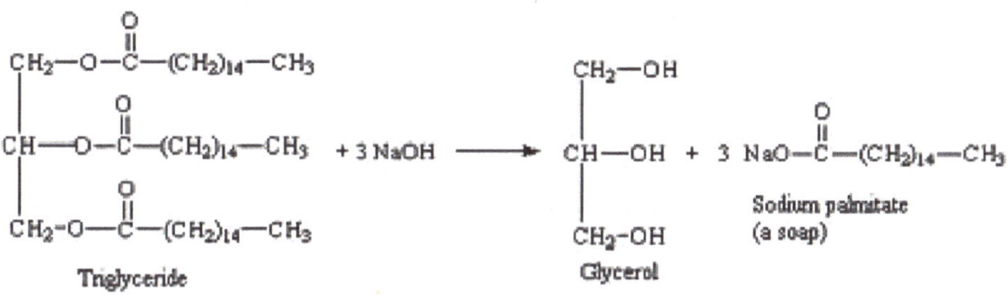

Hydrogenation: Hydrogenation refers to the addition of hydrogen to the double bonds of the fatty acids in acyl lipids like triacyl glycerols. The double bonds thus get saturated and the constituent fatty acids are converted to the appropriate saturated fatty acids.

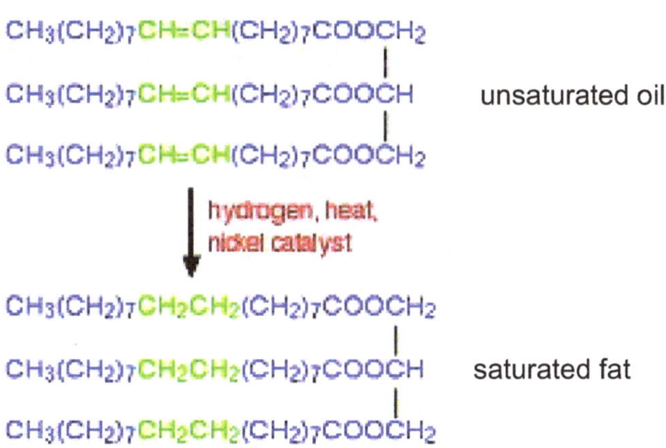

Hydrogenation of fats

The addition of hydrogen makes the unsaturated fatty acids saturated and the liquid oil gets converted to a solid fat. The reaction process is used industrially in the making of Vanaspathi (Dalda) from edible or even unconventional oils. Hydrogenated fat is a substitute for milk fat (butter oil). This synthetic fat has good shelf life similar to oils like sesame and groundnut oils. A small quantity of trans fats are formed during hydrogenation. The trans fats are considered unhealthy.

Saponification number: Saponification Number is defined as the number of milligrams of KOH required to saponify 1 g fat or oil. Saponification Number indicates the molecular weight or chain length of the fatty acids present. If the number is high, the molecular weight or chain length of the fatty acids is low and *vice versa*.

Iodine value (I V)/ iodine number: Fats and oils have double bonds in their fatty acids and hence respond to the addition of iodine. Iodine Number is defined as the number of grams of iodine absorbed by 100 g fat or oil. Iodine Number of fats / oils / acyl lipids indicates their degree of unsaturation. If the number is high, the degree of unsaturation is high i.e. more USFAs are present. Oils having less than 90 I V are non-drying oils. Those having 120 and above I V are called drying oils. Those with 90-120 I V are semidrying oils.

Acid value / number (A.V.): Acid Number/Value is defined as the number of milligrams of KOH required to neutralize the free fatty acids present in 1 g fat or oil. The free fatty acids may arise due to chemical or microbiological decomposition of the oils or fat during storage causing rancidity. Acid Number / Value is used to assess the degree of spoilage (rancidity) of a fat or oil. From the point of view of edible quality, an acid value of 2 or more is undesirable for the fat or oil. Fresh oils have low AVs while stored oils have high AVs. Antioxidants are added to prevent rancidity in oils.

Reichert-Meissel value (RM value): The number of ml of 0.1 N alkali required to neutralize the volatile fatty acids (separated after saponification through steam distillation) contained in 5 ml of fat is known as RM value. It is useful in quantifying the low molecular weight fatty acids in an oil.

Table 4. Analytical values of some fats/oils

Fat	Acid No.	RM Value	S.V.	I.V.
Human fat	-	0.40	196	65
Beef fat	0.25	-	198	40
Butter fat	0.40	26.0	220	27
Linseed oil	2.2	1.0	192	170
Coconut oil	1.5	7.0	258	8.0
Groundnut oil	0.6	-	191	94
Safflower oil	0.15	-	191	141
Castor oil	0.5	-	182	82

Lipid analysis

Lipid analysis is highly sensitive and sophisticated. It involves various steps. The initial step is extraction of lipids followed by elimination of non-lipid components. The pure lipids are then fractionated into individual classes using mainly thin layer chromatography(TLC). These individual lipids can be further analyzed by high performance liquid chromatography (HPLC) to identify the molecular species and structural elucidation. Triacylglycerols after saponification yield fatty acids. The FAs are converted to their methyl esters, which can be separated and identified with the help of gas liquid chromatography(GLC). GC-MS (gas chromatography- mass spectrometry) is more useful in structural elucidation. The reader may refer to advanced and specialized books for more information on lipid analysis.

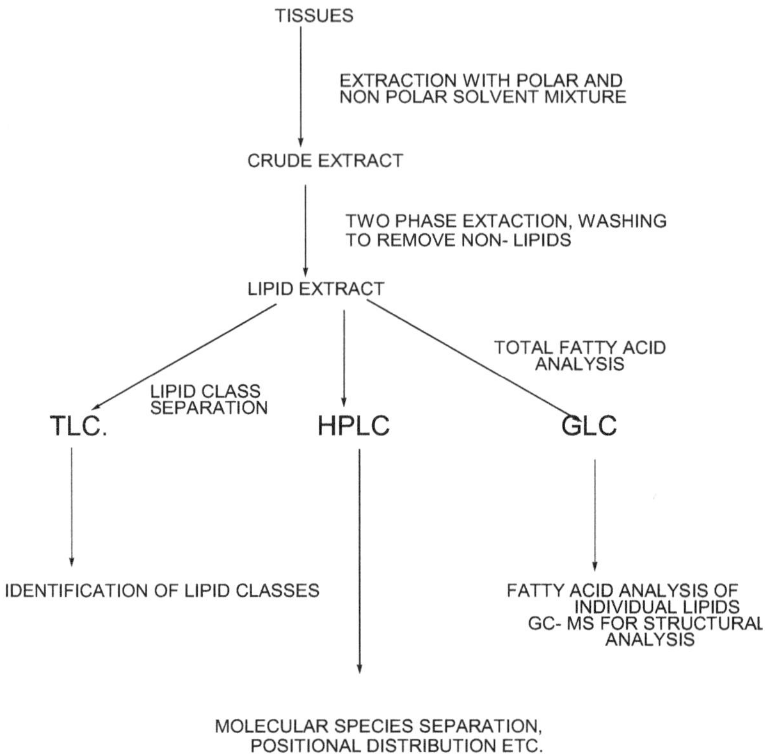

A SIMPLE FLOW CHART OF LIPID ANALYSIS

Industrial applications of acyl lipids

1. **Soaps and detergents:** Soaps and detergents are the most important cleansing agents in personal hygiene, hair care, dish washing and laundry. Soaps are the alkali metal salts of fatty acids. Since all of them contain an acyl group, acyl lipids mostly in the form of unconventional vegetable oils are used in the manufacture of soaps and detergents.

2. **Paints:** The paint industry uses acyl lipids in the form of vegetable oils as carriers of colorouring substances. Such oils are drying oils and the special feature is that coloured organic substances are readily dispersed / dissolved in such oils. After application to the surface, the oil gets oxidized by oxygen to form a thin film of solid layer keeping the colorizing matter also in the form of a thin film along with the oil. Among such oils, the linseed oil is the most important. Such oils used in making paints must be highly unsaturated.

3. **Rubber:** Natural rubber is obtained from the latex (milky secretion) of a tree *Hevea brasiliensis*. There are a few other species also. Rubber is highly elastic, water repellent, resistant to weak acids and alkalies, tough, impermeable, adhesive and insulator. These characteristics make rubber very useful in industries. Natural rubber in its latex form, however, as such cannot be used. To make it usable, it has to be vulcanized. In other words, the naturally occurring polyterpene rubber forms the basic material for the manufacture of better usable synthetic rubber.

4. **Biodiesel:** Biodiesel is a partial substitute for petroleum diesel and can be blended with the latter. Chemically, biodiesel is monoalkyl ester of fatty acids produced from vegetable oils by transesterification. In India, biodiesel is produced from the oils of Jatropha (wild castor, *Jatropha curcas*), Karanj (*Pongamia pinnata*) and a few other species. Edible oils like soybean oil, rape seed oil etc. are also used in U.S.A. and Europe. Biodiesel has some advantages, namely, (1) Easily biodegradable (2) Environmentally safe (3) Low fume emissions (4) Nontoxic and does not contain lead (5) Low green house gas emissions (6) The automobile engines need not be modified (7) The byproduct is glycerol, an important industrial chemical and (8) Calorific value is comparable to petroleum diesel. Major constraints in the use of biodiesel are 1. Availability of water for growing the crop, though water requirement is relatively less and 2. Availability of land.

Nutritional Role of Lipids

Most of the fat found in food is in the form of triglycerides, cholesterol, and phospholipids. Dietary fat facilitates in absorption of fat-soluble vitamins (A, D, E, and K) and carotenoids. Humans and other mammals have a dietary requirement for certain essential fatty acids, such as linoleic acid (an omega-6 fatty acid, ω-6 18:2) and alpha-linolenic acid (an omega-3 fatty acid, ω-3 18:3)) because they cannot be synthesized from simple precursors in the diet. Both of these fatty acids are 18-carbon polyunsaturated fatty acids differing in the number and position of the double bonds. Most vegetable oils are rich in linoleic acid (safflower, sunflower, and corn oils). Alpha-linolenic acid is found in the green leaves of plants, and in selected seeds, nuts, and legumes (linseed, rapeseed, walnut, and soy). Fish oils are rich in the longer-chain omega-3 fatty acids, eicosapentaenoic acid (EPA 20:5) and docosahexaenoic acid (DHA 22:6). Positive health benefits are associated with consumption of omega-3 fatty acids on infant development, cancer, cardiovascular diseases, and various mental illnesses, such as depression, attention-deficit hyperactivity disorder, and dementia. However, consumption of trans fats (similar in activity to SFAs), such as those present in partially hydrogenated vegetable oils, are a risk factor for cardiovascular disease.

Normally, fats and oils can form 20-30% of our food. Excessive dietary fat intake is linked to an increased risk of obesity and diabetes. In addition to the quantity, the fats consumed should include up to 1% of essential fatty acids listed above. In India the recommended level of visible dietary fat is 30g /day. Invisible fats through other foods like cereals, fruits, vegetables etc. account for up to 30g. From the point of view of good nutrition, instead of one single oil, two or three oils varying and containing all fatty acids should be consumed.

Chapter – 4

Amino Acids

Amino acids are the building blocks of proteins. They occur in all living plant cells as constituents of proteins and as free acids or amides. More than 200 naturally occurring amino acids have been identified in higher plants. Around 20 different amino acids have been observed to be incorporated into proteins. They have two functional groups, an amino group (-NH2) and a carboxylic group(-COOH). Both these groups are attached to the α carbon atom only. Amino acids with this structure are called α amino acids or 2-aminoacids. The unique characteristics of the 20 amino acids are due to the side chain (R), which can be an alkyl, hydroxyl, thiol, amino, sulphide, aromatic, or heterocyclic group.

Amino acids have the general formula, R-CH(NH_2)-COOH. The R represents a side chain. If R=H, the simplest amino acid glycine (NH_2 -CH_2 –COOH) results. Amino acids are alpha (α) amino carboxylic acids. The carbon atom is tetrahedral in shape. The various groups attached to it are placed in different positions. Since the valence of the carbon atom is four, four groups can be attached to the carbon atom. Based on the groups attached to the carbon atom it may be of two types.

1. **Symmetric carbon atom:** When the valence of the carbon is satisfied by more than one similar group, then the particular carbon atom is called symmetric carbon atom. Ex : Glycine. Compounds containing symmetric carbon atoms are optically inactive since they cannot rotate the plane of polarized light.
2. **Asymmetric carbon atom:** When the valence of a carbon atom is satisfied by four different groups, then that particular carbon atom is called as asymmetric carbon atom. Ex: Alanine. In the

asymmetric amino acid, to the α carbon atom, an amino group, a carboxylic group and a hydrogen atom are attached and the fourth one is the R group. This R group varies for each amino acid. All amino acids except glycine have at least one asymmetric carbon atom, hence they are all optically active.

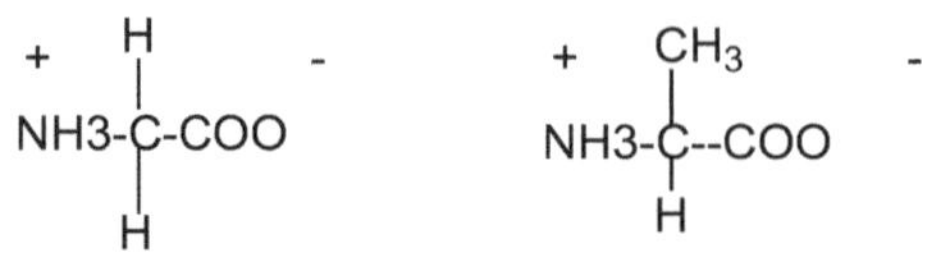

GLYCINE SYMMETRIC   ALANINE ASYMMETRIC CARBON

The asymmetric carbon exhibits chiral (means hand) character. The two isomers are non-superimposable and are called enantiomers. They can be distinguished on the basis of their ability to rotate plane polarized light differently. The common naturally occurring amino acids present in plants have the L-configuration, which means that they resemble L-glyceraldehyde.

In amino acids, the COOH group is acidic while the NH_2 group is basic. The groups interact, forming an inner salt, a zwitter ion (a compound with a cation and an anion as part of the same molecule), $^+H_3$ N-RCH-COO^-. It is also referred to as dipolar ion. Amino acids are crystalline solids at room temperature. They are more soluble in water than non-polar solvents due to their zwitter ionic character.

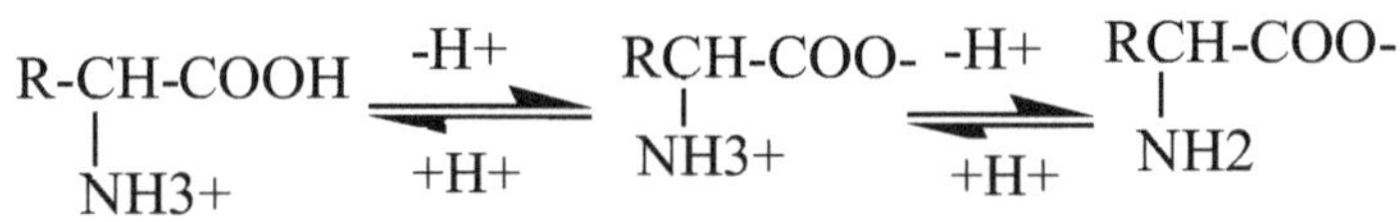

CATION   ZWITTER ION   ANION

The amino acids can be titrated with both acids and bases. The p^H at which it gives off equal number of anions and cations and does not show any migration when subjected to an electric field is known as the "isoelectric p^H". Above this p^H the amino acid gives off more negative ions and below this more positive ions. Aspartic acid has 2.77 and aspargine has 10.76 isoelectric p^H.

Table 1. Isoelectric points of some amino acids

Amino acid	Property	pH
Aspartic acid	Acidic	2.77
Glutamic acid	Acidic	3.22
Cysteine	Neutral	5.02
Methionine	Neutral	5.75
Tryptophan	Neutral	5.89
Glycine	Neutral	5.97
Alanine	Neutral	6.11
Lysine	Basic	9.74

Classification of amino acids: Amino acids can be classified in various ways.

1. **Based on side chains:** Based on the structure of the R groups, all the amino acids are classified as aliphatic, aromatic and heterocyclic amino acids.

2. **Based on their presence or absence in proteins:** Amino acids are classified as protein amino acids and non protein amino acids.

a) ***Protein amino acids:*** Amino acids that are used for synthesis of proteins are called protein amino acids. About 20 amino acids are present in proteins. The protein amino acids are listed below. Ornithine is normally not a protein amino acid.

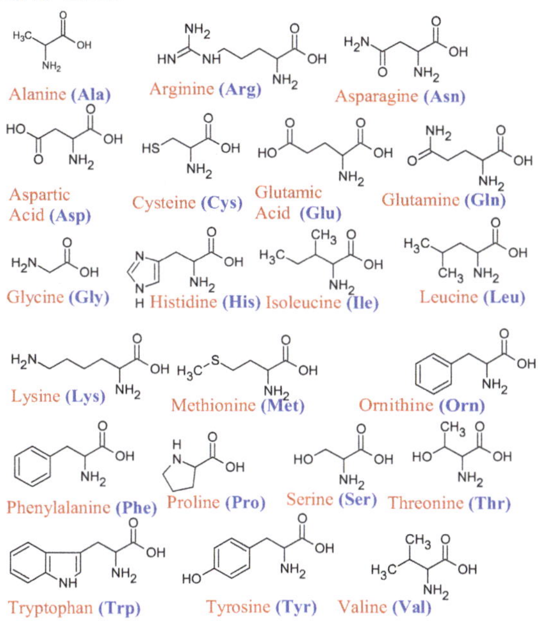

Protein Amino Acids

Protein aminoacids are often abbreviated to a three letter short form. For ex. arginine is abbreviated as Arg and glycine as gly. All the abbreviations are listed in the figure of aminoacids.

b) ***Non protein amino acids:*** Apart from the 20 amino acids that are present in proteins, several non protein amino acids are also present in nature. These are obtained by slight modification of 20 protein amino acids. Ex:- beta alanine, hydroxy proline, N- acetyl glutamic acid etc.

Amino acids can be simply classified into two groups, namely, polar and non-polar as follows:

Table 2. Polar and non-polar amino acids

Hydrophilic(polar)	Hydrophobic(non-polar)
Arginine (Arg)	Alanine (Ala)
Aspargine (Asn)	Isoleucine (Ile)
Aspartic acid(Asp)	Leucine (leu)
Cysteine (Cys)	Methionine(Met)
Glutamic acid (Glu)	Phenylalanine(Phe)
Glutamine (Gln)	Proline(Pro)
Histidine (His)	Tryptophan (Trp)
Lysine (Lys)	Tyrosine(Tyr)
Serine (Ser)	Valine (Val)
Threonine (Thr)	

A non-polar amino acid contains an alkyl or aromatic side chain. Hydrocarbons are non-polar and not soluble in water and hence are hydrophobic (water hating). Polar amino acids have side chains that contain polar groups like hydroxyl (-OH), thiol(-SH), amino ($-NH_2$), and heterocyclic amines. The electronegative atoms in the side chains make these amino acids hydrophilic (water loving) by forming hydrogen bonds with water.

All the amino acids occur in almost all proteins. The acidic amino acids aspartic acid and glutamic acids are rich in cereal proteins. The basic amino acids lysine and hydroxyl lysine occur in collagen, the connective tissue.

3. **Based on requirement to the body as essential and non essential:** Animals cannot synthesise all the 20 amino acids that are present in proteins. Some have to be provided to the body through the external diet. The amino acids which cannot be synthesized by the body and thus have to be supplied through diet are called essential amino acids. On the other hand, some amino acids can be synthesized by the body, and they are called as non-essential amino acids.

Essential amino acids

Body tissue contains 20 amino acids. Out of this arginine, histidine, isoleucine, leucine, lysine, methionine, phenylalanine, threonine, tryptophan and valine are called essential amino acids. They cannot be synthesized by the human body. Hence they have to be supplemented through the diet.

The quality or balance of a protein depends upon the types and amounts of different amino acids present in it. Availability of the essential amino acids mostly accounts for this quality. Ruminants such as cattle and sheep are able to utilize protein synthesized by the rumen microorganisms. Hence the diet of ruminants is of little consequence. Animal proteins are more balanced in essential amino acids than plant proteins. Cereals are deficient in lysine, methionine, tryptophan and threonine. Pulse proteins are deficient mostly in methionine. Oilseed proteins are deficient in lysine, methionine and threonine. Non-ruminants such as man, pigs and poultry need to be careful with respect to the availability of essential amino acids in their diet.

Table 3. Essential and non essential aminoacids

Essential aminoacids	Non-essential amono acids
Methionine	Alanine
Arginine	Aspartic acid
Threonine	Glutamine
Tryptophan	Cysteine
Valine	Glycine
Isoleucine	Proline
Leucine	Serine
Phenyl alanine	Tyrosine
Lysine	
Histidine (for children only)	

Properties of amino acids

Amino acids with non-polar R groups being hydrophobic tend to fold themselves and have a compact structure. For Ex. proline and hydroxyl proline have a unique structure. Amino acids with polar R groups are hydrophilic and tend to ionize in aqeous solutions. For Ex. aspartic and glutamic acids have a net negative charge at neutral $P^{H.}$

All amino acids except glycine have an asymmetric carbon atom and hence exist in L- and D- configurations. A second asymmetric carbon atom is present in four amino acids, namely, isoleucine, hydroxyl lysine, threonine and hydroxyl proline.

Due to the presence of an amino group and a carboxylic group the amino acids possess dipolar and zwitterion characteristics. They are called amphoteric. They are zwitterionic at physiological p^H of 6-7. Amino acids act as both acids and bases. In fact they act as buffers in living beings. At low p^H values the amino acid is protonated to –COOH and the amino acid becomes positively charged. At higher p^H values the amino group is deprotonated to $-NH_2$ and the amino acid becomes negatively charged.

Table 4. Charge of an amino acid at different p^H values

$^+NH_3$ H C R COOH	$^+NH_3$ H C R COO$^-$	NH_2 H C R COO$^-$
Cation p^H 1	Zwitter ion p^H 7	Anion p^H 11

Reactions of amino acids

1. Amino acids can form esters with alcohols and salts with bases.

2. They form amides with NH_3. Aspargine and glutamine transport NH_3 in the body.

3. The amino group can be methylated or benzylated.

4. Nitrous acid reacts with the amino group liberating N_2 and the corresponding hydroxyl acid.

$$RCHNH_2COOH + OHNO \rightarrow RCHOHCOOH + N_2 + H_2O$$

This reaction is the basis for the Van Slyke method for determining the number of free amino groups in a protein.

5. Formaldehyde reacts with amino group to form a methylene compound

$$NH_2RCHCOOH + HCHO \rightarrow CH_2{=}NRCHCOOH + H_2O$$

The resulting acid can be titrated to estimate the amino acids. This is called "Formal titration". The amino acids react with other aldehydes to form Schiffs bases.

6. *Ninhydrin test*: Ninhydrin is an oxidizing agent which oxidatively de-aminates the alpha-amino groups of amino acids. It is very important for the detection and the quantitative analysis of amino acids. Ninhydrin

also reacts with primary amines. However the formation of carbon dioxide is quite diagnostic for amino acids. Alpha amino acids yield a purple substance (Ruhemann's purple) that absorbs maximally at 570 nm. Imino acids (proline, hydroxyproline) yield a yellow product.

When a solution of amino acid is heated with ninhydrin, the amino acid is oxidatively de-aminated to produce ammonia and a ketoacid. The keto acid is decarboxylated to produce an aldehyde with one carbon atom less than the parent amino acid. The net reaction is that, ninhydrin oxidatively de-aminates and decarboxylates α-amino acids to CO2, NH3 and an aldehyde. The reduced ninhydrin then reacts with the liberated ammonia and another molecule of intact ninhydrin to produce a purple colored compound known as Ruhemann's purple. This ninhydrin reaction is employed in the quantitative determination of amino acids. Proteins and peptides that have free amino groups (in the side chain) will also react and give color with ninhydrin.

7. Amino acids react with dinitro fluorobenzene, phenyl iso thiocyanate and dansyl chloride, which are also useful in their quantitative estimation.

8. Tyrosine, tryptophan and phenyl alanine, the aromatic amino acids absorb at 280nm. Most proteins contain tyrosine and this property is used for estimation of proteins by U.V. spectrophotometry.

Separation of amino acids

Amino acids can be separated based on their isoelectric points using a chromatographic technique called electrophoresis. A bufferd amino acid mixture is applied to a gel on a thin plate or piece of filter paper, which is connected to two electrodes. A voltage applied to the electrodes causes the positively charged amino acids to move towards cathode while the negatively charged amino acids move towards the anode. Amino acids at the isoelectric point with zero charge do not move. Sufficient time is allowed for the amino acids to move under the electric field and the gel or paper is removed. Ninhydrin is sprayed to visualize the amino acids. They are identified based on their direction and movement (Rf values – movement relative to the solvent front). They can be recovered from the gel or paper by cutting out the spots and extracting them individually into suitable solvents. The identities can be further confirmed using spectroscopic techniques.

For example, we can separate a mixture of amino acids, namely, lysine, valine and aspartic acid in a buffer at p^H 6.0. Aspartic acid moves towards the anode, lysine moves towards the cathode and valine, being a zwitter ion at p^H 6.0, does not move and remains at the place of spotting. Electrophoresis is a widely used technique for separation of charged compounds.

Chapter – 5

Peptides

An amine reacts with a carboxylic acid to form an amide. Similarly, the amino group on one amino acid molecule reacts with the carboxyl group of another amino acid to form a compound called peptide. A peptide bond is an amide bond that forms when the $-COO^-$ group of one amino acid reacts with the $-NH_3^+$ group of another amino acid. The resulting compounds are called peptides. Even after this there is a reactive amino group and a carboxyl group in the peptide. If these groups react further a poly peptide can form and later a polymer called protein can form.

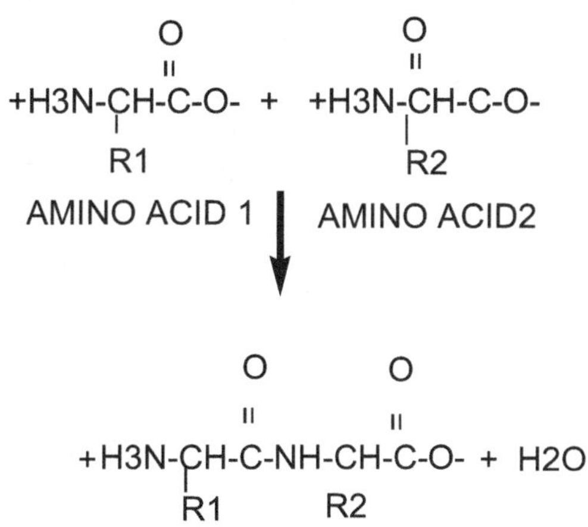

A PEPTIDE

Peptide bond formation: Amino acids are linked together by formation of covalent bonds. The covalent bond is formed between the α-carboxyl group of one amino acid and the α-amino group of the next amino acid. The bond so formed between the carboxyl and the amino groups, after elimination of a water molecule is called as a peptide bond and the compound formed is a peptide.

A peptide is a chain of amino acids linked together by peptide bonds. The peptide bond has a partial double bond character and is mostly in the trans- configuration. Proteins on partial hydrolysis yield the polypeptides and oligopeptides. When only two amino acids are joined together, the product is called a dipeptide and when three amino acids are joined, the product is a tripeptide. Polypeptides are usually long peptides (10-100 amino acids) whereas oligopeptides are short (< 10 amino acids). Proteins are made up of one or more polypeptides with more than 50 amino acids.

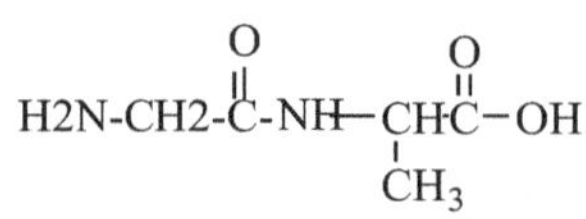

GLYCYLALANINE- A DIPEPTIDE

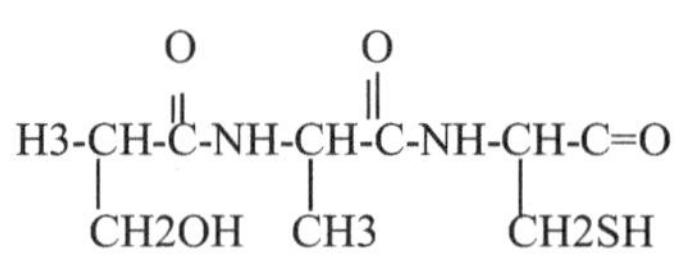

SERYLALANYLCYSEINE-A TRIPEPTIDE

Peptides have both free amino and free carboxyl groups, similar to the amino acids. Hence they exhibit reactions similar to that of amino acids. Peptides give a reaction called "biuret "reaction. Treatment of a peptide or a protein with Cu^{2+} in an alkaline solution yields a purple reaction which has absorption at 540nm. This property is used to estimate proteins.

The peptide bond between the carbon and nitrogen exhibits partial double bond character due to closeness of the carbonyl carbon-oxygen double-bond allowing the resonance structures. The peptide unit which is made up of the CO-NH atoms is thus relatively rigid and planar, permitting adjacent peptide units to be at different angles.

$$-O=C-NH \longleftrightarrow -\overset{-}{O}C=\overset{+}{N}H-$$

RESONANCE STRUCTURES

Nomenclature of the peptides

In a peptide, always the first amino acid has its N terminal free. It will not be involved in the formation of a peptide bond, but the carboxyl group of first amino acid and amino group of second amino acid are involved in the formation of the peptide bond. While naming a peptide, the first amino acid is usually named by adding yl and the second amino acid as it is. Ex. Glycyl alanine. A tripeptide consisting of alanine, glycine and serine is named as alanylglycylserine.

Classification of peptides based on structure

Peptides can be classified based on the structure as linear or cyclic peptides and based on the number of amino acids involved in the formation of the peptide.

1. **Linear peptide:** When a peptide has a linear structure it is called as a linear peptide. Ex. Glutathione and insulin. Number of peptide bonds present in a linear peptide = (n – 1) where n represents the number of amino acids.

a) **Glutathione:** It is a small molecule made up of three amino acids, which exists in almost every cell of the body. This is called as natural redox tripeptide. It is present in all cells. It plays an important role in plant respiration. It functions to keep enzymes active in animal cells. It also serves as a source of – SH groups in cells. There is an unusual peptide bond present between glutamic acid and cysteine and glycine. It is chemically called as gamma glutamyl cysteinyl glycine.

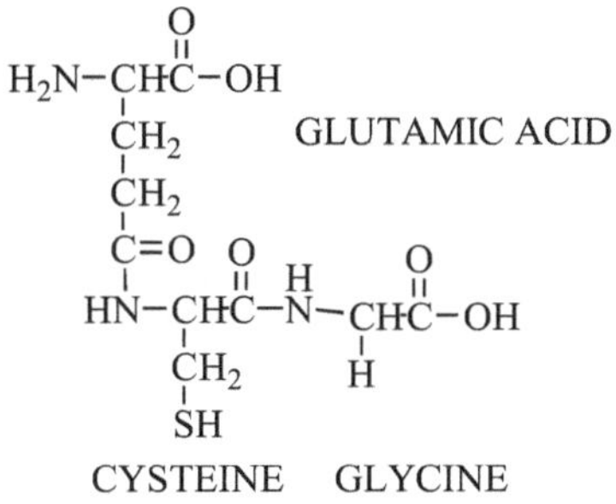

GLUTATHIONE - A TRIPEPTIDE

b) ***Insulin:*** The peptide hormone insulin is produced by clusters of specialized cells called the beta cells of islets of Langerhans of pancreas. It is synthesized as large precursor molecule pre-proinsulin. Pre-proinsulin undergoes partial hydrolysis and forms proinsulin with the removal of signal peptide. Proinsulin is made up of three chains, chain A, B and C. In proinsulin, chain A and C are connected by chain B which is made up of 30 amino acids. It undergoes a proteolytic cleavage and forms insulin with the removal of chain B. Chain A and chain C are connected by means of two disulphide bonds in insulin molecule. The molecular weight of insulin is 5700.

Insulin increases the usage of glucose and increases the storage of glycogen. Insulin secretion by the pancreas is stimulated by a rise in the level blood sugar. The glucose content of the blood gets reduced with an increase of glycogen in the liver. The insulin used in medicine is isolated from the pancreatic tissues of slaughter house animals.

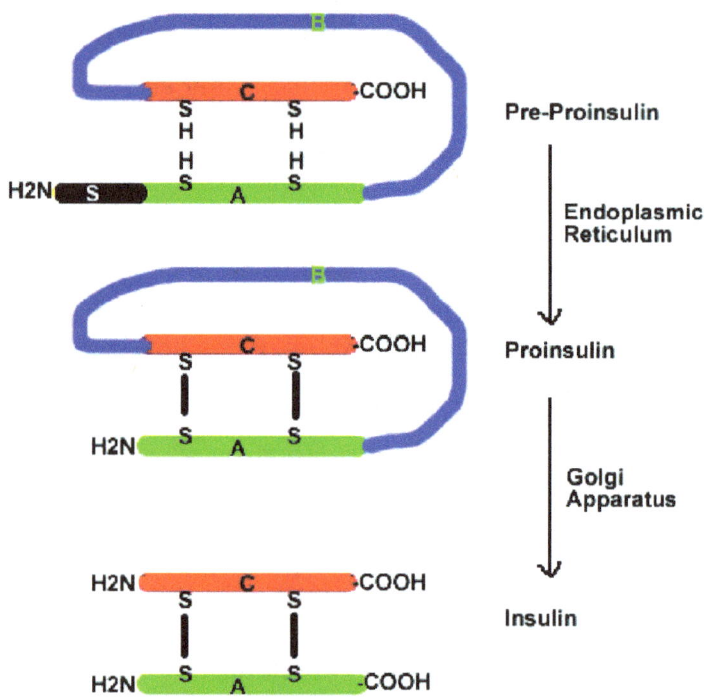

Cyclic peptides: If the carboxyl group at the C-terminus of a peptide forms a peptide bond with the N-terminal amino group, a cyclic peptide is formed. Cyclic peptide has its structure in the form of a ring. The number of peptide bonds in a cyclic peptide can be calculated by the formula: No of peptide bonds = n, where n= no of amino acids present in a peptide. This is in contrast to n-1 peptide bonds in a linear peptide.

Cyclic peptides are most commonly found in microorganisms, and often incorporate some D-amino acids as well as unusual amino acids like ornithine (Orn).

a) ***Gramicidin-S and Bradykinin:*** Gramicidin-S, an antibiotic cyclic peptide is produced by a strain of *Bacillus brevis*. Gramicidine – S is a decapeptide made up of two pentapeptides. It has non -protein amino acids such as D-Phenyl alanine, which is synthesized by the microbes. Bradykinin, a nonapeptide is produced by the cleavage of a larger protein in the blood. The amino acid sequence in it is arg-pro-pro-gly-phe-ser-pro-phe-arg. It is similar to prostaglandin and reduces B.P. It also stimulates smooth muscle tissues and increases capillary permeability.

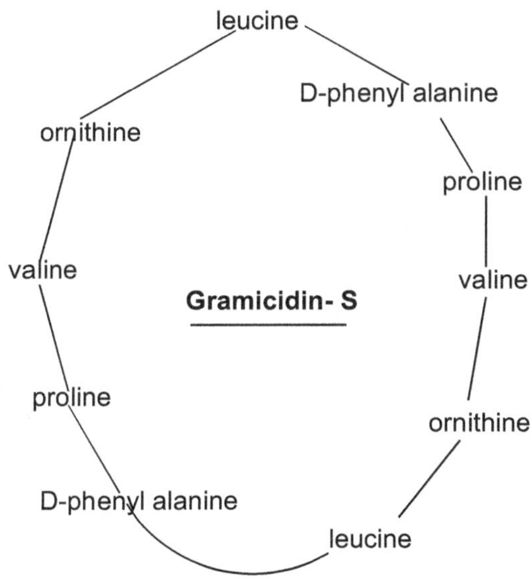

b) ***Vasopressin and oxytocin:*** These two cyclic peptides have been isolated from the pituitary gland. They have cyclic structures with a cystine unit. The two differ with a few amino acids. Oxytocin has isoleucine, tyrosine, cystine, aspartic acid and glutamic acid in the ring structure and proline, leucine, and glycine in the side chain. Vasopressin has phenyl alanine, tyrosine, cystine, aspartic acid and glutamic acid in the ring structure and proline, arginine and glutamic acid in the side chain.

Vasopressin acts to increase the amount of water reabsorbed by the kidneys. Hence it is called anti-diuretic hormone (ADH). It also increases B.P. and hence used to treat low B.P. Oxytocin is involved in smooth muscle contraction. It stimulates the uterus and has a role in child birth and lactation.

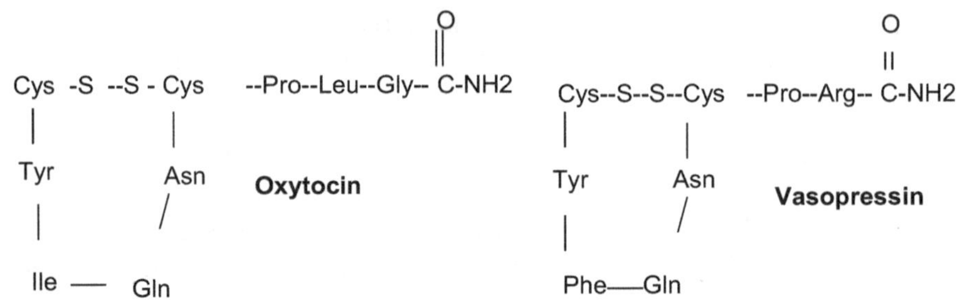

Table 5. Effects of oxytocin and vaso pressin

Function /property	Vasopressin	Oxytocin
Water diuresis	Inhibits	No effect
Blood pressure	Elevates	Lowers minimally
Coronary arteries	Constricts	Dilates limitedly
Intestinal contractions	Stimulates	No effect
Uterine contractions	Stimulates	Stimulates
Ejection of milk	Stimulates slightly	Stimulates

c) ***Malformin:*** Malformin is another example of a cyclic peptide. Source is mycelium of fungus *Aspergillus niger*. Malformin promotes cell elongation, causes several malformations on the stem and petioles of several plant species and also causes root curvatures. Chemically it is cyclo-D-cysteinyl-D-cysteinyl-L-valinyl-D-leucinyl-L-isoleucine.

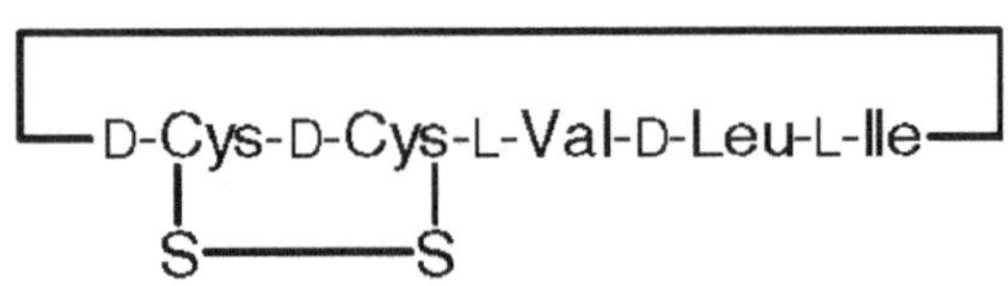

Chapter – 6

Proteins

The word protein was first coined in 1838 to emphasize the importance of this class of molecules. The word is derived from the Greek word proteios which means "of the first- rank". Proteins are polymers/macro molecules made up of several amino acids. When there are more than 50 amino acids in a chain, the polypeptide is normally called a protein. Each protein in the cells of living beings has a unique sequence of amino acids that determines its biological function. The proteins are folded into specific defined structures, which are maintained by a large number of relatively weak bonds. The three dimensional structure is well defined and suited to it's specific function.

Very small changes in the structure can modify the function. Hence, it is important to know the structure of proteins to understand their behavior and function. The biological activity of proteins depends on maintenance of folded conformation. Proteins fold into well defined three dimensional shapes and they are able to recognize their corresponding substrates or antigen molecules and bind them tightly. The protein structure has been classified into four different levels based on the folded conformation of the protein.

Primary structure: Primary structure is the simplest level of structural organization. The sequence of the different amino acids in a protein is called the primary structure of the peptide or protein. Though it appears simple, it is highly important. The conformation and function of a protein are determined by the primary structure. Even a small change in one amino acid may seriously affect the biological activity of a protein. For example, a hormone that stimulates the thyroid to release thyroxine consists of a tripeptide glutamine-histidine-prolamine. Although other sequences are possible for these three amino acids, only the tripeptide with the Glu-His-Pro sequence of amino acids has hormonal activity.

Similarly, hemoglobin is made up of four polypeptide chains, two chains of α type and two chains of β type. When glutamic acid present at the 6th position from N terminal end in the β chain is replaced by valine, Hemoglobin-S is formed and causes sickle cell anemia by which there is a reduction in capacity to carry oxygen by hemoglobin. In a polypeptide, numbering of residues always starts at the N terminal end (NH2-group), where the amino group is not involved in a peptide bond formation.

The unique sequence of amino acids is responsible for many of the fundamental properties of a protein. Proteins also can have circular structure, namely," bite their own tail", example: the K carrier, valinomycin. With 20 naturally occurring L-amino acids, it is possible to have a large number of combinations and permutations. For a dipeptide there are $20^2 = 400$ combinations, for a tripeptide, $20^3 = 8000$ and for 100 amino acids, it is $20^{100} = 1.27 \times 10^{130}$ combinations.

Secondary structure: The secondary structure of protein refers to the conformational patterns of the polypeptide chain. An extended peptide folds itself to be stable. The three dimensional arrangement of the protein is referred to as the secondary structure. Different types of secondary structures occur widely in proteins. The most prominent are the α helix and β conformations. Linus Pauling and Robert Corey predicted the existence of these secondary structures in 1951. In general, proteins fold into two broad classes of structures, namely globular proteins and fibrous proteins. Globular proteins are compactly folded and coiled, whereas, fibrous proteins are more filamentous or elongated.

The α-Helix: The α-helix is a common secondary structure encountered in proteins of the globular class. The formation of the α-helix is spontaneous and is stabilized by H-bonding between amide nitrogen and carbonyl carbons of peptide bonds spaced four residues apart.

Features of α-Helix: a) In this, the polypeptide backbone is tightly wound around an imaginary axis drawn longitudinally through the middle of the helix and R groups protrude outward from the helical back bone. b) Single turn of the helix has 5.4 amino acids and it is called as the pitch of the helix. c) There are 3.6 amino acid residues per turn of the helix. d) Distance between the peptide bonds is 1.5 A^0. e) The helix structure is maintained by the hydrogen bond formed between every first and the fourth amino acids of the helix. f) The hydrogen bonds are intra molecular and are parallel to the central axis.

There are three types of α- helices based on the direction and the nature i.e.: left handed α- helix, right handed α- helix and triple helix Ex. collagen has β - Pleated Sheets: An α-helix is composed of a single linear array of helically disposed amino acids and β-sheets are composed of 2 or more different regions of stretches of at least 5-10 amino acids. The folding and alignment of stretches of the polypeptide backbone aside one another to form β-sheets is stabilized by hydrogen bonding between amide nitrogen and carbonyl carbons. However, the hydrogen bonding residues are present in adjacently opposed stretches of the polypeptide backbone as opposed to a linearly continuous region of the backbone in the α-helix.

β-sheets are said to be pleated. This is due to positioning of the α-carbons of the peptide bond which alternates above and below the plane of the sheet. β-sheets are either parallel or antiparallel. In parallel sheets, adjacent peptide chains proceed in the same direction (i.e. the direction of N-terminal to C-terminal ends is the same), whereas, in antiparallel sheets adjacent chains are aligned in opposite directions. β-sheets can be depicted in ball and stick format or as ribbons in certain protein formats.

Tertiary structure: The tertiary structure refers to the complete three-dimensional structure of the polypeptide units of a given protein. It is the spatial relationship of different secondary structures to one another within a polypeptide chain and how these secondary structures themselves fold into the three-dimensional form of the protein. The tertiary structure is maintained by different forces. These include hydrogen bonding, hydrophobic interactions, electrostatic interactions and Van der Waals forces.

Hydrogen Bonding: Polypeptides contain numerous proton donors and acceptors both in their backbone and in the R-groups of the amino acids. The environment in which proteins are found also contains ample H-bond donors and acceptors of the water molecule. H-bonding, therefore, occurs not only within and between polypeptide chains but with the surrounding aqueous medium.

Hydrophobic forces: Proteins are composed of amino acids that contain either hydrophilic or hydrophobic R-groups. It is the nature of the interaction of the different R-groups with the aqueous environment that plays a major role in shaping protein structure. The hydrophobicity of certain amino acid R-groups tends to drive them away from the exterior of proteins into the interior. This driving force restricts the available conformations into which a protein may fold.

Electrostatic forces: Electrostatic forces refer to the interaction of ionized R groups of amino acids with the dipole of the water molecule. The slight dipole moment that exists in the polar R-groups of amino acid also influences their interaction with water. It is, therefore, understandable that the majority of the amino acids found on the exterior surfaces of globular proteins contain charged or polar R-groups.

Van der Waals forces: There are both attractive and repulsive Van der Waals forces that control protein folding. Attractive Van der Waals forces involve the interactions among induced dipoles that arise from fluctuations in the charge densities that occur between adjacent uncharged non-bonded atoms. Repulsive Van der Waals forces involve the interactions that occur when uncharged non-bonded atoms come very close together but do not induce dipoles. The repulsion is the result of the electron-electron repulsion that occurs as two clouds of electrons begin to overlap. Although Van der Waals forces are extremely weak, relative to other forces governing conformation, it is the huge number of such interactions that occur in large protein molecules that make them significant to the folding of proteins.

Quaternary Structure: Many proteins contain 2 or more different polypeptide chains that are held in association by the same non-covalent forces that stabilize the tertiary structures of proteins. Proteins with multiple polypetide chains are oligomeric proteins. The structure formed by monomer-monomer interaction in an oligomeric protein is known as quaternary structure. Oligomeric proteins can be composed of multiple identical polypeptide chains or multiple distinct polypeptide chains. Proteins with identical subunits are termed homo-oligomers. Ex: Acetylcholine receptor. Proteins containing several distinct polypeptide chains are termed hetero-oligomers. Hemoglobin, the oxygen carrying protein of the blood, contains two α and two β subunits arranged with a quaternary structure in the form, $\alpha 2\ \beta 2$. Hemoglobin is, therefore, a hetero-oligomeric protein. Sulfide bridges also occur in the quaternary structure of proteins. They can occur within the chain or between the chains.

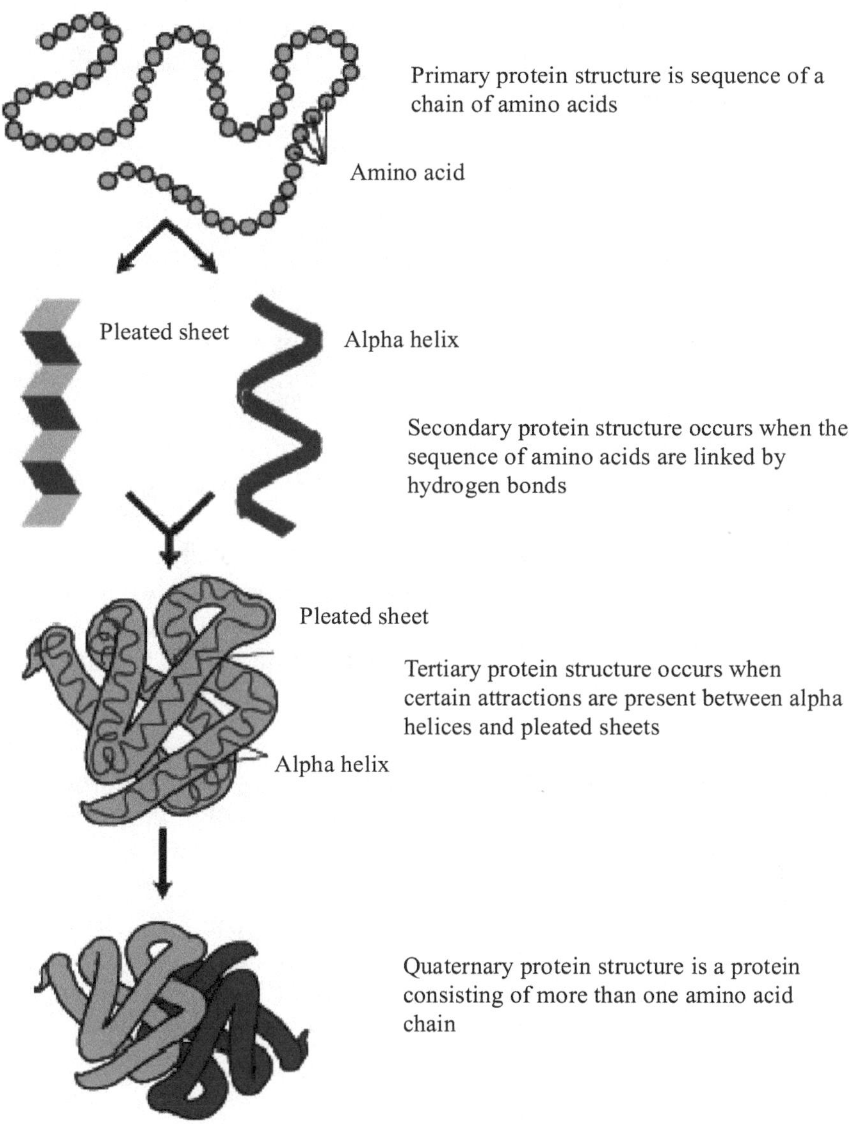

Levels of organization of protein sturcture

Properties of proteins

1. **U.V absorption:** Proteins absorb U.V. light at 280 nm due to the presence of aromatic amino acids like tryptophan and tyrosine. This property is used in the estimation of proteins.

2. **Isoelectric point:** Isoelectric point is also called as isoelectric pH. This is the pH at which the number of positive and negative charges are equal in the protein and they are electrically neutral. Solubility of proteins is least at the isoelectric pH.

3. **Zwitterions:** Proteins contain both positive and negative charges and hence they are called as zwitterions. Amino acids act as zwitterions as they can donate a proton and form cations. They can as well accept a proton and form an anion. Each amino acid can act as an anion, cation, neutral species and as zwitterion.

4. **Immunological properties:** Proteins exhibit a special property called immunological property, which is useful in defense mechanism. Whenever any antigen(foreign macromolecule) enters the body, it releases a special class of proteins called antibodies (immunoglobulins). The interaction of antigen and antibody to form the antigen-anti body complex is called an immune reaction. Antigen/immunogen may be a protein (protein coat of virus), or a carbohydrate (sugars on the bacterial outer coat) or nucleic acid. Antibodies are special glycoproteins which recognize and bind antigens specifically. The anti body immobilizes the antibody after binding to it. During malnutrition, the immunity is low.

   The immune system consists of two types, namely, cellular immune system and humoral immune system. The cellular immune system is mediated by Tcells, Thymus derived lymphocytes and are active against microorganisms, fungi, and parasites. The humoral immune system is mediated by the 'B' cells, which secrete antibodies. The immune system has memory, in addition to specificity. A vaccine contains a weakened form of an antigen. Hence it allows the body to build up sufficient defense against the antibody. A few important immunoglobulins are listed in the Table 1.

Table 1. Some important immunoglbulins

Ig type	Hchain (heavy)	L chain (light)	Sugars	Occurrence
Major Ig G	Gamma	Kappa/lambda	4%	Plasma
Ig alpha	Alpha	Kappa/lambda	10%	Colostrum
Ig M	Mew	Kappa/lambda	15%	Plasma
Minor Ig D	Delta	Kappa/lambda	18%	—
Ig E	Eta	Kappa/lambda	18%	Active against allergies

5. **Denaturation:** It is a physical change in which there is a collapse of the protein structure. Due to denaturation, there is a decrease in solubility and loss of biological activity of proteins. Denaturation occurs at extreme temperatures and pH and also by many chemicals like organic solvents, urea, ionic detergents etc. Denaturation is due to disruption of bonds that stabilize the secondary, tertiary, or quaternary structure of proteins. On denaturation, non covalent bonds in the protein are broken and its primary structure remains intact. When favorable conditions arise, some peoteins, namely, the reversibly denatured proteins, spontaneously return back to its native biologically active form and this process is called renaturation.

   Denaturing agents are mostly heat, acids and bases, organic compounds, heavy metal ions and mechanical agitation. Heat breaks down hydrogen bonds. Cooking is a denaturation process which however, does not affect the nutritional properties of food. Lactic acid production by bacteria, is utilized in preparation of curd, which is also an example of denaturation. Alcohol swab denatures proteins in bacteria at the time of injection of medicines. Heavy metal ions when ingested denature enzymes in our body and act as toxins. The whipping of eggs is an example of mechanical action of denaturation.

6. **Protein folding:** Many proteins fold to their native conformation on their own by self assembly. However several other accessory proteins help in bends. They include enzymes and molecular chaperons.

   a) *Enzymes:* Ex: Peptidyl prolyl cis-trans isomerase introduces reverse bends.

   b) *Molecular chaperons:* They are a special class of proteins which help in the folding of other proteins. Molecular chaperons identify the improperly folded proteins and provide a microenvironment in which a polypeptide can progressively fold itself. Molecular chaperons do not impose a structure to proteins but only provide the required environment to the protein. They belong to heat shock protein family which protects a polypeptide from denaturation and aggregation at high temperatures.

7. **Solubility:** Protein solubility is influenced by pH, heavy metals, salts and organic solvents.

   a) *pH:* Solubility of proteins is minimum at isoelectric point. At acidic pH, proteins behave like cations. Anion forms of some compounds like Trichloroacetic acid are effective in precipitating the proteins. Solubility is influenced by the presence of polar hydrophilic groups on the surface.

   b) *Heavy metals:* At alkaline pH, proteins behave like anions. Cationic forms of some heavy metals such as Mercury and Lead are attracted by the negative charges present on the free side chains of proteins and precipitate them.

   c) *Salts:* The addition of a small quantity of salt like ammonium sulphate or sodium chloride, increases the solubility of protein due to increased ionic strength of the solution. Further increase of salt concentration, decreases the solubility of the protein.

   d) *Organic solvents:* Organic solvents like ethanol, acetone and butanol lower the dielectric constant of the medium and decreases the solubility of proteins.

Protein hydrolysis: Proteins are hydrolyzed to obtain individual amino acids. This process occurs in the stomach when enzymes like pepsin and trypsin catalyze the hydrolysis of proteins.

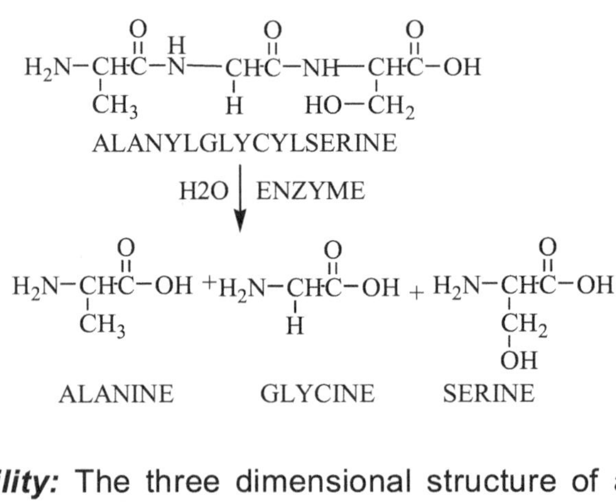

Protein stability: The three dimensional structure of a protein is maintained by a range of noncovalent interactions, namely, electrostatic force, hydrogen bond, hydrophobic force and disulphide bonds in addition to the peptide bonds between amino acids.

Elcetrostatic forces: These are electrostatic interactions between two ionic groups of opposite charge. Example, the NH_3 group of Lys and the -C=O group of Asp, often referred to as an ion pair or salt bridge.

Hydrogen bonds: These are electrostatic interactions between weakly acidic group and a partially negative charge that attracts the hydrogen atom. Hydrogen bonds are stronger than Van der Walls forces, but weaker than covalent bonds. Hydrogen bonds play an important role in protein structure and in the double helix structure of DNA and lipid bilayers. Hydrogen bonds are critical to the properties of water and its role as a biochemical solvent.

Hydrophobic forces: Nonpolar molecules minimize their contact with water. This is what we call hydrophobic effect. Proteins find a conformation in which their nonpolar side chains are out of contact with H_2O. Hydrophobic forces are important in determining the protein structure folding and stability.

Disulphide bonds: These covalent bonds are formed between Cys residues that are closer together in a protein. The disulphide bonds help stabilize the three dimentional structure of a protein. Disulphide bonds are present in extracellular and secreted proteins

Sequencing of amino acids

There are several effective methods by which a polypeptide end groups may be identified. The most effective method in identification of N-terminal residue is Edman degradation method named after its inventor Edman. Phenylisothiocynate reacts with the N- terminal amino groups of proteins under mildly alkaline conditions to form their phenylthiocarbamyl derivative. This product on reaction with an anhydrous strong acid such as trifluoroacetic acid (TFA, F3CCOOH), forms thiazolinone derivative with the cleavage of a peptide bond involving the carboxyl group of the N terminal residue. The treatment with TFA does not affect the other peptide bonds leaving a peptide chain with n-1 amino acid residues. The Edman degradation therefore releases the N- terminal amino acid residue but leaves intact the rest of the polypeptide chain.

The treatment of the thiazolinone derivative with aqueous acid leads to the formation of Phenylthiohydantoin (PTH) derivative of N-terminal amino acid which can be identified by chromatography. The residual peptide chain can be now submitted to a new coupling reaction and the next amino acid can be identified.

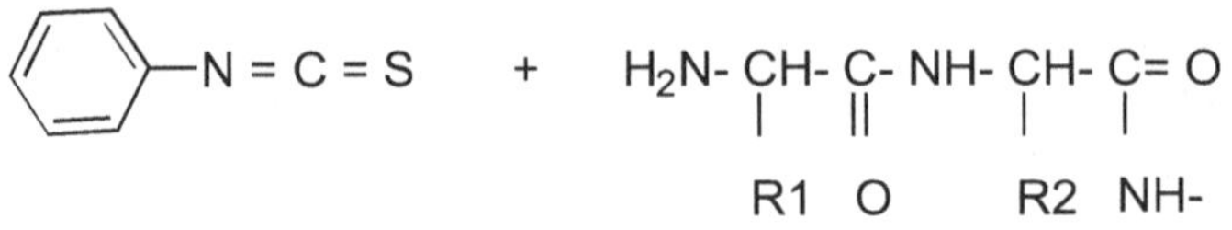

Phenyl isothio- cyanate

Peptide

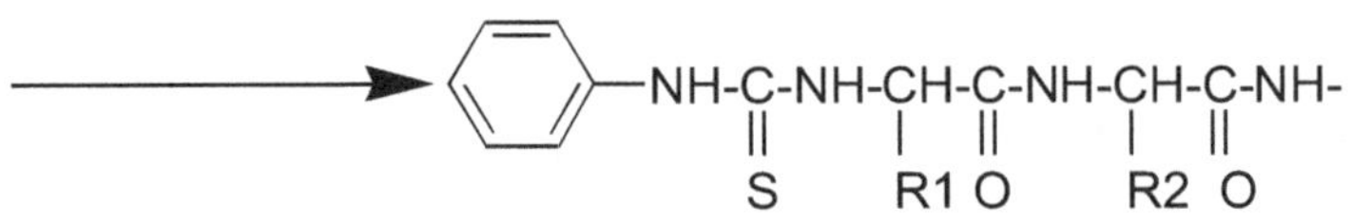

Phenyl isothio carbamyl peptide

Purification techniques

The aim of protein purification is to isolate a specific protein from all the others in a mixture of proteins. A combination of fractionation techniques is used for this purpose. The properties based on which the proteins are separated are the size, charge on the molecule and the specific binding ability. The following techniques are used for purification of proteins.

a) ***Salting in and salting out:*** Increase in the solubility of a protein by addition of small quantities of sodium chloride is called as salting in and this is due to an increase in the ionic strength of the solution. On the other hand, when excess of salt is added to the solution, there is decrease in the solubility and this is called as salting out. Salting out occurs due to hydrophobic effect. Both the hydrophobic and hydrophilic amino acids are present in protein molecules. The hydrophobic amino acids generally are present in the interior of the protein but some are present on the surface also in small patches. Water molecules become ordered, when they are forced to interact with these patches.

When the salt concentration is increased, a competition develops for the water between the protein and the salt. Some of the water molecules are attracted by the salt ions, which decreases the number of water molecules available to interact with the charged part of the protein. As the salt concentration increases, the water on protein is removed thus exposing the hydrophobic area of protein molecule. These hydrophobic areas on the protein molecule get attracted to each other by hydrophobic effect. This results in increase in weight of the molecule and its aggregation. Larger the surface hydrophobic area on a protein molecule, quicker will be the precipitation of the protein at a lower concentration of the salt.

For example, 0.8 M ammonium sulfate precipitates out the fibrinogen protein from the blood serum, whereas 2.4 M ammonium sulfate is required to precipitate the albumins. Salting out is also used to concentrate the proteins from a dilute solution. The precipitated protein is re-dissolved in a small quantity of solvent.

b) ***Dialysis:*** Proteins can be separated by dialysis through a semi permeable membrane such as cellulose membrane which has pores in it. Bigger molecules are retained in the bag and smaller molecules pass through the membranes. As the size of protein molecules are above 10kDa, this technique is not so effective. Further, at equilibrium the quantity of small molecule inside and outside the bag are equal. Several charges are needed to reduce the quantity of small molecules.

c) ***Gel filtration or Size Exclusion Chromatography:*** This chromatographic technique is based upon the use of a porous gel in the form of insoluble beads placed into a column. When a solution of proteins is passed through the column, a protein of small size is likely to enter the pores. Small proteins can penetrate into the pores of the beads and, therefore, are retarded in their rate of travel through the column. The larger proteins will move through the gaps present in the column and are likely to move fast and are collected first when compared to the smaller ones. Different beads with different pore sizes can be used depending upon the desired protein size separation profile. Polymer beads made up of dextrose, agarose or polyacrylamide are generally used and the commercial names for few polymer beads are Sephadex, Sepharose and Biogel respectively.

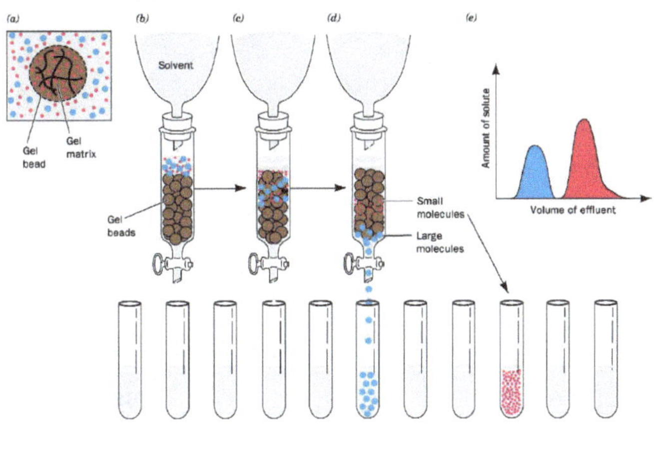

d) ***Ion Exchange Chromatography:*** Each individual protein exhibits a distinct overall net charge at a given pH. Some proteins will be negatively charged and some will be positively charged at the same pH. This property of proteins is the basis for ion exchange chromatography. The basic principle is, like charges repel and unlike charges attract each other. Fine cellulose resins are used that are either negatively (cation exchanger) or positively (anion exchanger) charged. Proteins of opposite charge to the resin are retained as a solution of proteins is passed through the column. The bound proteins are then eluted by passing a solution of ions bearing a charge opposite to that of the column. By utilizing a gradient of increasing ionic strength, proteins with increasing affinity for the resin are progressively eluted. CM (carboxy methyl)-Cellulose and DEAE(diethyl aminoethyl)-Cellulose are examples of negatively charged and positively charged resins.

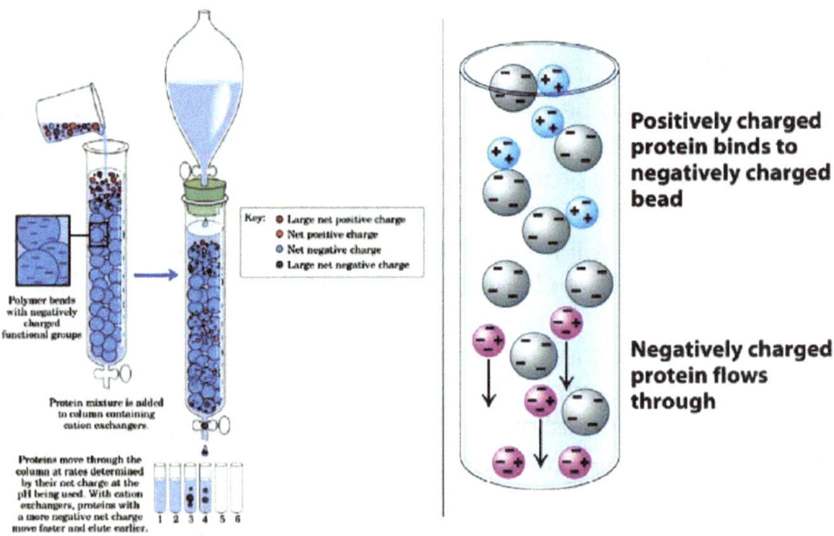

Ion Exchange Chromatography

DEAE- Cellulose is positively charged and is called an anion exchanger. CM -cellulose is negatively charged and acts as a cation exchanger. Proteins can be bound to the resin and can be displaced by increasing the strength of buffer and or changing the p^H. Proteins are eluted from a column in order of least strongly bound to the most strongly bound.

Classification of proteins

Proteins can be classified into two general classes, namely, simple and conjugated proteins. Simple proteins on hydrolysis yield only amino acids while conjugated proteins yield additionally an organic or inorganic group called prosthetic group. The simple proteins are further classified into albumins, globulins, glutelins, prolamines, histones, protamines and albuminoids.

Most enzymes and harmone proteins belong to either albumins or globulins. Plant seed proteins belong to glutelins or globulins. Proteins can also be classified into two classes namely, globular and fibrous proteins. Globular proteins have spherical compact structures. They have very specific interaction with other molecules. Globular proteins are made up of both hydrophobic and hydrophilic amino acids. In these proteins, hydrophobic amino acids are present internally. Enzymes are examples of globular proteins. They have storage and transport functions. Fibrous proteins have a high tensile and mechanical strength. They provide structural support. Fibrous proteins which are insoluble proteins are made up of mostly hydrophobic amino acids. Examples of fibrous proteins are collagen, keratin, silk, wool etc.

Table 2. Properties of different proteins

Name	Soluble in water	Soluble in salt solution	Soluble in acid/alkali	Occurence
Albumins	Soluble	Soluble		Milk,eggs, plant & animal cells
Globulins		Soluble		Milk,eggs,meat &seed proteins
Glutelins			Soluble	Cereals & plant material
Prolamines (sol in 80% alcohol)				Cereals & plant material
Histones	Soluble		Soluble	Glandular tissues(organs)
Protamine	Soluble		Soluble	Fish sperm
Albuminoids				Meat

Proteins are also classified on the basis of their functions.

a) ***Catalytic proteins:*** These are enzyme proteins that catalyze chemical and biochemical reactions within the living cell and outside. This group of proteins probably is the biggest and most important group of the proteins. Enzymes are responsible for all metabolic reactions in the living cells. Ex: DNA and RNA polymerases, dehydrogenases, trypsin, etc.

b) ***Regulatory proteins:*** *Some* hormones are examples of this class of proteins that are responsible for the regulation of many processes in organisms. Ex: Insulin, corticotropin, repressors, etc.

c) ***Transport proteins:*** These proteins are involved in transporting some chemical compounds and ions. Ex: Haemoglobin, myoglobin, serum albumin, etc.

d) ***Defense proteins:*** These proteins are involved in the defense mechanism of the cell. Ex: Gamma globulins, antibodies, fibrinogen, thrombin, toxins, venom, ricin .

e) ***Structural proteins:*** These proteins are involved in maintaining the structure of other biological components like cells and tissues. Ex: Collagen, elastin, keratin, fibroin, proteoglycans, etc.

f) ***Contractile/motile proteins:*** These proteins are involved in contraction of the tissues. Ex: Actin and myosin are responsible for muscular motion. Tubulin and dynein are the other examples.

g) ***Storage/nutrient proteins:*** These proteins contain energy, which can be released during various metabolic processes in the organism. Ex: casein(milk), gliadin(wheat), ovalbumin(egg), ferritin, etc.

h) ***Receptor proteins:*** These proteins act as receptor molecules. They are responsible for signal detection and translation into other type of signal. Ex: GTPases.

Protein quality evaluation methods

Proteins are vital to the living processes and carry out a wide range of functions essential for the sustenance of life. In judging the adequacy of dietary proteins to meet the human needs, not only the quantity, but also the nutritional quality of the dietary protein is important. Proteins present in different foods vary in their nutritional quality because of differences in their amino acid composition. The human body derives the essential amino acids from the dietary protein. The quality of the dietary protein depends on its essential amino acid composition and its digestibility. The best quality protein is the one which provides essential amino acid pattern very close to the pattern of reference proteins such as egg protein or milk protein. The quality of a protein is evaluated by the following methods.

Biological methods: Proteins of raw grains (particularly legumes) are less digestible than that of animal foods. Generally, the low digestibility of plant proteins is due to the presence of trypsin inhibitors which are destroyed on cooking. The overall quality of a protein can be determined by biological methods with laboratory animals like rats as follows;

a) *Protein efficiency ratio (PER):* The gain in the weight of young animals per unit weight of protein consumed is measured and the value obtained is designated as PER. Protein efficiency ratio is defined as gain in body weight by protein intake.

$$PER = \frac{gain\ in\ body\ weight\,(g)}{protein\ intake\,(g)}$$

b) *Digestibility coefficient (DC):* Dietary proteins are hydrolyzed to amino acids on digestion by various proteolytic enzymes. Proteins differ in the digestibility depending on their source. Since only amino acids are absorbed into the blood stream, undigested portion of the protein is excreted in the fecal matter. The term digestibility coefficient of protein refers to the percentage of the ingested protein absorbed into the blood stream after the process of digestion is complete. When an animal is fed on nitrogen free diet, certain amount of nitrogen is excreted in the fecal matter. This is derived mainly from the digestive juices. This is called endogenous fecal nitrogen. When a protein food is given, the nitrogen found in feces consists of both endogenous nitrogen and food nitrogen lost in digestion. To find out nitrogen lost in digestion, endogenous faecal nitrogen should also be determined. For the determination of the digestibility coefficient, the data required are

1. Food nitrogen intake – *In*
2. Total fecal nitrogen excreted – *Fn*
3. Endogenous fecal nitrogen – *Fe*

$$DC = \frac{In-(Fn-Fe)}{In} x100$$

Where Fn – Fe is the food nitrogen lost in digestion.

The digestibility coefficient of proteins is influenced by several factors, such as indigestible carbohydrates and proteolytic enzyme inhibitors etc.

c) *Biological value (BV):* The amount of nitrogen (N) in the diet and in excreta of adult animals is measured and the percentage of N retained by animals from out of N absorbed from the diet is calculated. The value thus obtained is the biological value of the protein. It measures the quantity of dietary protein utilized by the animal for meeting its protein needs for sustenance and growth.

1. Food nitrogen intake - *In*
2. Total fecal nitrogen excreted - *Fn*
3. Endogenous fecal nitrogen - *Fe*
4. Total urinary nitrogen excreted - *Un*
5. Endogenous urinary nitrogen - *Ue*

$$B\,V = \frac{gain\ in\ body\ weight\,(g)}{protein\ intake\,(g)}$$

$$B\,V = \frac{In - (Fn - Fe) - Un - Ue}{In - (Fn - Fe)} \times 100$$

Chapter – 7

Enzymes

Enzymes are organic catalysts produced by living organisms. They make possible biochemical reactions. Hence they are called biocatalysts. They are mostly a specialized class of proteins. However, there are some non-protein enzymes like ribozymes, which are RNAses. Enzymes speed up reactions at ambient temperatures. This is in contrast to chemical reactions like saponification of lipids which take place when lipids are boiled for a few hours with concentrated alkali. Enzymes, namely, lipases, hydrolyze lipids at body temperatures in minutes. They are superior to chemical catalysts by a factor of 10^7 to 10^{14}. An enzyme may be a simple protein or a complex protein. They are produced by cells and are mostly globular proteins. Each cell contains an estimated 3000 different enzymes.

The names of enzymes describe the compound or the reaction that is catalyzed. The actual names of enzymes are derived by replacing the name of the reaction or the reacting compound with the suffix *ase*. For example, an oxidase catalyzes an oxidation reaction, and a dehydrogenase removes hydrogen atoms. Sucrase hydrolyses sucrose to glucose and fructose while lipases hydrolyse lipids to fatty acids and glycerol. In olden nomenclature the names of enzymes end with the suffix *in* for example papain pepsin, rennin and trypsin.

Characteristics of enzymes

1. Enzymes are biocatalysts. Hence they are not destroyed during an enzyme action. In other words, they are regenerated at the end of the reaction.
2. They can only speed up the reaction but cannot initiate any reaction. They speed up the reaction by 10^7 to 10^{14} times. These reactions are slow in the absence of enzymes. Time factor is very critical for biological reactions. Hence enzymes play a vital role.
3. Enzymes do not alter the equilibrium constant of a reaction but alter the rate at which equilibrium is reached. A non- enzymatic reaction may take several years to reach equilibrium, while an enzymatic reaction may take a second or less. Both forward and reverse reactions can take place at the same speed.
4. Enzymes are very specific. Their specificity with regard to substrate and the reaction that they catalyze are well known.
   a. Enzymes produce only those products that are needed by the organism.
   b. Enzymes permit cells to use energy and materials efficiently depending on cellular needs.

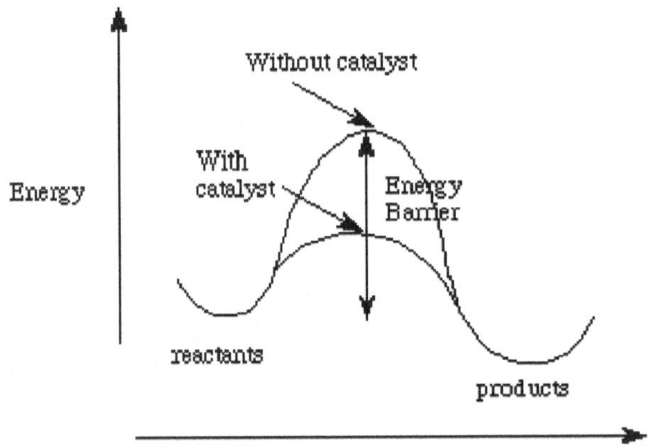

Catalyst lowers energy barrier, reaction goes faster

Chemical nature of enzyme

J.B.Sumner first proposed that enzymes are protein in nature. Many enzymes require the presence of other compounds such as cofactors, coenzymes and metal ions for their catalytic activity. This entire active complex is referred to as the holoenzyme; i.e., apoenzyme (protein portion) plus the cofactor (coenzyme, prosthetic group or metal ion or activator). Many enzymes are conjugated proteins in which the prostate group has the actual catalytic activity. Apoenzyme + Cofactor = Holoenzyme. Apoenzyme is a protein substance which is thermo labile. The cofactor can be: 1. A coenzyme - a substance which is thermo-stable and loosely attached to the protein part. 2. A prosthetic group - an organic substance which is thermo-stable and is firmly attached to the protein. 3. A metal ion/ activators like K^{+}, Fe^{++}, Fe^{+++}, Cu^{++}, Co^{++}, Zn^{++}, Mn^{++}, Mg^{++}, Ca^{++}, and Mo^{+++}. Many enzymes have quaternary structure. They consist of two or more polypeptide chains, each with secondary and tertiary structures. Sometimes the structures are identical, but the functions as catalyst may be different.

Specificity of Enzymes

Enzymes are very important as diagnostic and research tools due to the specificity they exhibit relative to the reactions they catalyze. A few enzymes exhibit absolute specificity; that is, they catalyze only one particular reaction. Other enzymes are specific for a particular type of chemical bond or functional group. In general, there are four distinct types of specificity.

1. ***Absolute specificity:*** The enzyme catalyzes only one reaction by acting on a single type of substrate. Ex: Glucokinase converts only glucose to glucose 6-phosphate. Urease catalyzes the hydrolysis of only urea.
2. **Group specificity:** The enzyme acts on a group of related molecules. Ex : Hexokinase acts on all hexoses and converts them to their hexose phosphates.
3. **Linkage specificity:** The enzyme acts on a particular type of chemical bond regardless of the rest of the molecular structure. Ex: Proteases act on various proteins. Chymotrypsin catalyzes the hydrolysis of peptide bonds.

4. **Stereochemical specificity:** The enzyme acts on a particular steric or optical isomer. Ex: L-Amino acid oxidases act only on L- Amino acids.

Active site

It is the location of an enzyme at which place the substrates bind and undergo a chemical reaction. The active site of an enzyme is usually found in a cleft or pocket that is lined by amino acid residues (or nucleotides in ribozymes) that participates in recognition of the substrate. Residues that directly participate in the catalytic reaction mechanism are called active site residues.

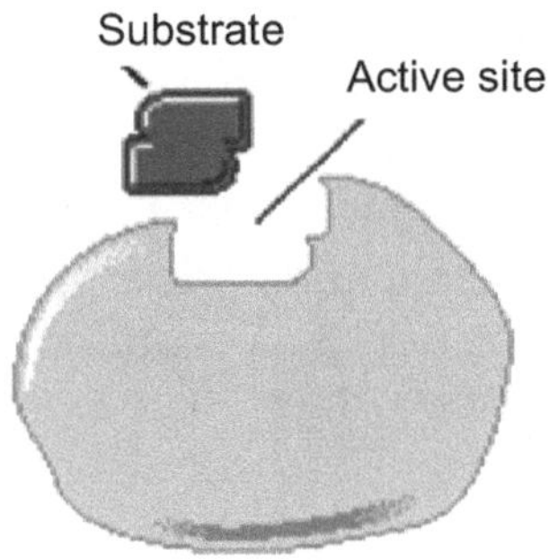

Enzyme action is often explained by the lock and key analogy. The substrate must fit a portion of the enzyme called "active site" similar to the lock and key system. Within the active site the amino acid side chains take part in catalyzing the reaction.

Mode of enzyme action

A chemical reaction such as A → P takes place because a certain fraction of the substrate possesses enough energy to attain an activated condition called the transition state. The minimum energy required by the substrate to cross the energy barrier and thereby form products is called the activation energy. Every reactant has the required amount of energy but during the course of time some energy is lost when the reactants undergo collisions in the form of heat. Hence enzymes decrease the activation energy so that more number of reactants cross the energy barrier and products are formed. The transition state of substrate is at the top of the energy barrier separating the reactants and products. The rate of a given chemical reaction is proportional to the concentration of this transition state species. The energy of activation is the amount of energy required to bring all the molecules in 1 mole of a substance at a given temperature to the transition state. Enzymes combine transiently with the substrate to produce a transition state intermediate having a lower energy of activation than the uncatalysed reaction. Thus they accelerate chemical reactions by lowering the energy of activation.

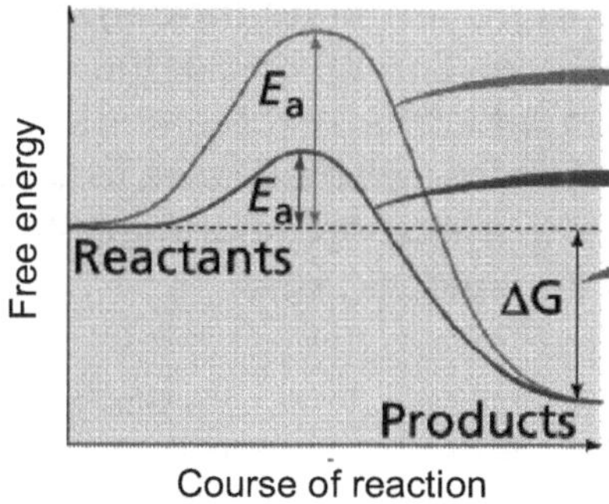

An **uncatalyzed reaction** requires a higher activation energy than does a **catalyzed reaction**

There is no difference in free energy between catalyzed and uncatalyzed reactions

Example: $H2O2 \xrightarrow{\text{CATALASE}} H2O + (O)$

Table 1. Activation energies of different reactions

Reaction condition:	Activation energy (KCal mol-1)
Uncatalyzed reaction	18
Catalyzed by catalase	7

It is generally believed that the catalytic reactions occur in at least two steps. Step 1: A molecule of enzyme (E) and a molecule of substrate (S) collide and react to form an intermediate called the enzyme-substrate complex (ES). Step 2: The decomposition of ES complex to give product (P) and the active enzyme. The formation of an ES complex favors lower activation energy.

$$S + E \longrightarrow (ES) \longrightarrow P + E$$

SUCROSE + SUCRASE ⟶ SUCROSE-SUCRASE COMPLEX
⟶ SUCRASE + GLUCOSE + FRUCTOSE

Two theories have been proposed to explain the mechanism of enzyme action.

1.**Fischer's lock and key theory (Rigid template model):** According to this theory proposed by Emil Fischer during 1890s, the active site possesses a unique conformation which is complementary to the structure of the substrate thus enabling the two molecules to fit together in much the same way as a key fits into a lock. An unfortunate feature of this model is the implied rigidity of the catalytic site.

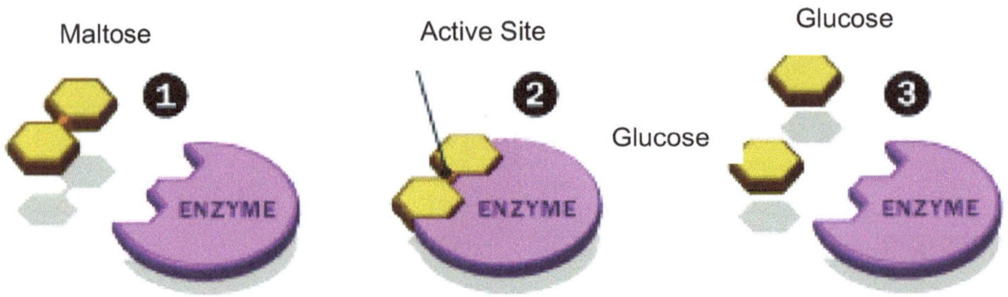

2. **Koshland's induced-fit theory:** Koshland advocated a theory to account for the specificity of enzymes. He postulated that the essential functional groups on the active site of the free enzyme are not in their optimal positions for promoting catalysis. When the molecule is bound by the enzyme the catalytic groups assume favorable geometrical positions to form the transition state. The enzyme molecule is unstable in this active conformation and tends to revert to its free form in the absence of substrate. In the induced fit model, the substrate induces a conformational change in the enzyme which aligns the amino acid residues/ other groups for substrate binding, catalysis or both.

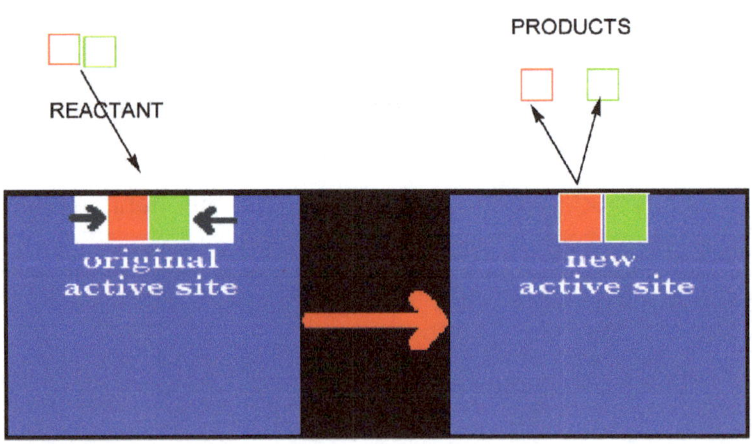

Measurement of enzyme activity

The presence or absence of an enzyme is typically determined by observing the rate of the reaction the enzyme catalyzes. The goal of most enzyme assays is to quantitatively measure the amount of enzyme activity present in a sample. Thus, assay results are typically reported in “activity” units. A unit of activity may be defined in various ways. But all such units are ultimately based on rates of substrate consumption and/or product formation. The most common units for expressing catalytic activity are the International Unit (U) and the Katal (Kat). The International Union of Biochemistry (IUB) originally defined a standard unit of enzyme activity (1 U) as that amount of enzyme that catalyzes the formation of 1 μmol product (or the conversion of 1 μmol substrate) per minute under standard conditions. The specific activity is the number of enzyme units per mg of protein.

In 1979, the Nomenclature Committee of the IUB recommended the use of the Katal as the fundamental unit of enzyme activity. The Katal is defined as that amount of enzyme that catalyzes the formation of 1 mol product (or the conversion of 1 mol substrate) per second under defined conditions. 1 U = 1/60 micro katal = 16.67 nano katal.

The turn over number of an enzyme is the number of substrate molecules transformed per unit time by a single enzyme molecule. The enzyme carbonic anhydrase has a turnover number of 36,000,000 per minute per enzyme molecule. Present in red blood cells, it catalyzes the following reaction:

$$CO_2 + H_2O \rightarrow H_2CO_3$$

The H_2CO_3 formed in tissues is transported to lungs where it is exhaled as CO_2.

Factors affecting enzyme activity

The factors that mainly influence any enzyme-catalyzed reaction are 1. Substrate concentration, 2. Enzyme concentration, 3. Temperature, 4. pH, and 5. Inhibitors. Other factors such as state of enzyme (oxidation), time and activators also affect enzyme-catalyzed reaction to certain extent.

1. **Substrate Concentration:** It has been shown experimentally that if the amount of an enzyme is kept constant and the substrate concentration is gradually increased, the reaction velocity will increase until it reaches a maximum. After this point, increases in substrate concentration will not increase the velocity. Keeping the factors such as pH, temperature and enzyme concentration at optimum levels, if the substrate concentration [S] is increased, the velocity of the reaction also increases to a certain extent. At very low substrate concentration, the initial reaction velocity (V) is nearly proportional to the substrate concentration (first order kinetics). However, if the substrate concentration is increased further, the rate of increase slows down (mixed order kinetics). With more increase in the substrate concentration the reaction rate approaches a constant (zero order-reaction where velocity is independent of substrate concentration).

At the initial point, even though the substrate molecules are present in excess over the enzyme on molar basis, not all the enzyme molecules present combine with the substrate. Hence, increasing the substrate concentration will increase the amount of enzyme associated with substrate as ES and thus V will depend on [S]. At Vmax, all the enzyme molecules are saturated with substrate molecules so that further increase in [S] cannot result in increased reaction rate. Michaelis-Menten derived an equation to explain this type of behavior.

$$V = (Vmax(S)) / (Km + (S))$$

At half maximal velocity [S] = Km

$$\text{i.e. } Vmax / 2 = (Vmax(S)) / (Km + (S))$$

$$Km + S/2 = Vmax(S)/Vmax$$

$$Km + S = 2S$$

$$Km = 2S - S = S$$

$$Km = S$$

Hence, Michaelis - Menten constant (Km) is defined as the substrate concentration at half maximal velocity and is expressed as mole per litre. *Significance of Km:* A small Km indicates that the enzyme requires only a small amount of substrate to become saturated. Hence, the maximum velocity is reached at relatively low substrate concentrations. A large Km indicates the need for high substrate concentrations to achieve maximum reaction velocity. The substrate with the lowest Km upon which the enzyme acts as a catalyst is frequently assumed to be enzyme's natural substrate, though this is not true for all enzymes.

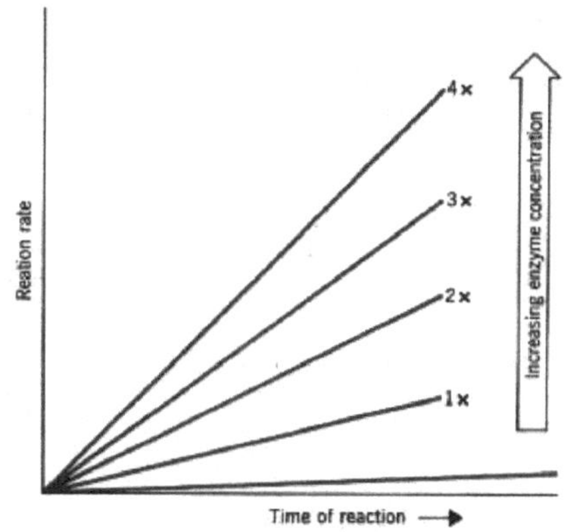

2. **Enzyme Concentration:** When compared to substrate concentration, the concentration of enzyme is always very low on molar basis. Hence, increasing the enzyme concentration will always increase the reaction rate until the substrate becomes a limiting factor. In order to study the effect of increasing the enzyme concentration upon the reaction rate, the substrate must be present in an excess amount; i.e., the reaction must be independent of the substrate concentration. Any change in the amount of product formed over a specified period of time will be dependent upon the level of enzyme present. When a purified enzyme is used the rate of reaction is proportional to the concentration of the enzyme over a fairly wide range.

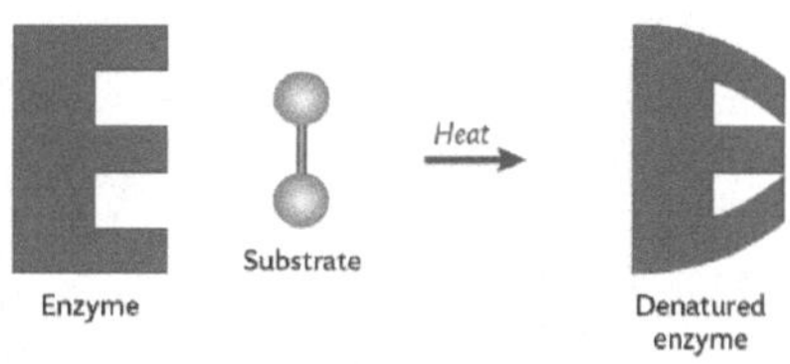

3. **Temperature Effect:** Like most chemical reactions, the rate of an enzyme catalyzed reaction increases as the temperature is raised, because enzymes are inactive below their optimum temperature. The reaction rate increases with temperature to a maximum level, then abruptly declines with further increase in temperature, because most enzymes rapidly become denatured at temperatures above 40°C. Over a limited range of temperature (10-50^{0}C), the velocity of enzyme-catalyzed reactions roughly double or even treble with a 10^{0}C rise in temperature. At higher temperatures of 50^{0}C, the higher enzyme activity is short lived due to denaturation process that sets in simultaneously.

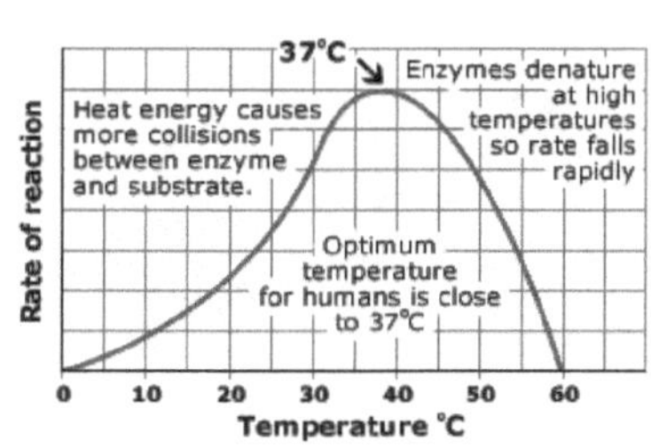

4. **Effect of pH:** Enzymes are affected by changes in pH. The most favorable pH value, the point where the enzyme is most active is known

as the optimum pH. Extremely high or low pH values generally result in complete loss of activity for most enzymes. pH is also a factor in the stability of enzymes. For each enzyme, there is also a region of pH

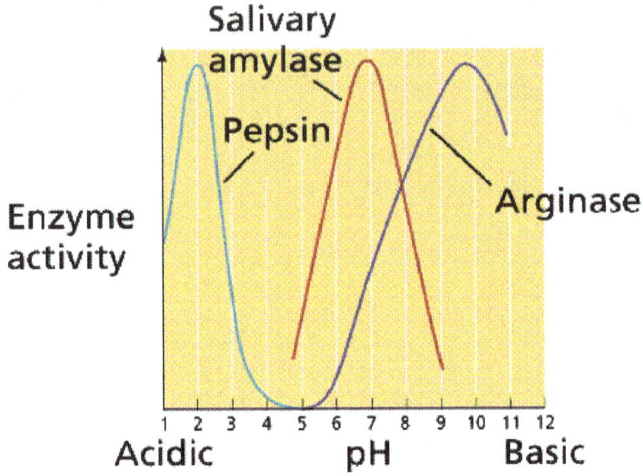

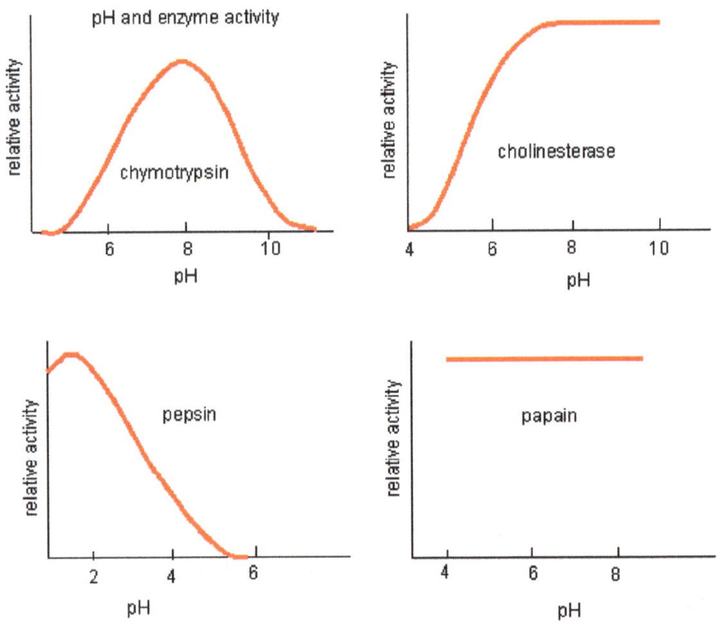

optimal stability. The optimum pH value will vary greatly from one enzyme to another. Large variations in p^H denatures the enzyme.

5. **Effect of end products:** If the concentration of end products increase, the enzyme activity gets reduced (slow reaction). Some products, if

they are acid or alkaline, may alter the p^H of the system and affect the enzyme activity. The effect of the end product on the enzyme activity is sometimes called chemical feed back or negative feed back.

6. **Other factors:** In addition to temperature and pH there are other factors, such as ionic strength, which can affect the enzymatic reaction. Ions activate the enzymes by a number of methods. They become part of the active site. They may change the equilibrium constant. Na+, K+, Mg++, Ca++,Zn++, Mn++, Al+++, etc. activate, while Hg++,Pb ++ or Ag+ poison the enzymes. Each of these physical and chemical parameters must be considered and optimized for an enzymatic reaction to be accurate and reproducible.

Enzymes undergo physical changes during the reaction but revert to their original form at the end of the reaction. Enzymes exhibit enormous catalytic power. Ex: Carbonic anhydrase enzyme catalyses the conversion of carbon dioxide to carbonic acid. CO2 + H2O $\rightarrow$ H2CO3. In this reaction, each enzyme molecule can hydrate 10^5 molecules of carbon dioxide per second. Enzyme activity is regulated in a variety of ways, ranging from control over the amount of enzyme protein synthesized by the cell or modulation of activity through reversible interaction with metabolic inhibitors and activators or through isoenzymes.

Cofactors

A large number of enzymes require an additional non-protein component to carry out their catalytic functions. Generally these non-protein components are called as cofactors. The cofactors may be either one or more inorganic ions such as Fe^{++}, Mg^{++}, Mn^{++} and Zn^{++} or complex organic molecules called coenzymes. Some enzymes require both coenzyme and one or more metal ions for their activity. Metals are required as cofactors in approximately two thirds of all enzymes. Metallo-enzymes contain a definite quantity of functional metal ion that is retained throughout. Metal-activated enzymes bind metals less tightly but require added metals. The distinction between metallo-enzymes and metal activated enzymes thus rests on the affinity of a particular enzyme for its metal ion. The iron-sulfur enzymes are a unique class of metalloenzymes in which the active centre consists of one or more clusters of sulfur-bridged iron chelates. These are of greater importance in plant systems.

Inhibitors

An enzyme inhibitor is a molecule that binds to enzymes and decreases their activity. Since blocking an enzyme's activity can kill a pathogen or correct a metabolic imbalance, many drugs are enzyme inhibitors. They are also used as herbicides and pesticides. The binding of an inhibitor can stop a substrate from entering the enzyme's active site and/or hinder the enzyme from catalyzing its reaction. Inhibitor binding is either reversible or irreversible. Irreversible inhibitors usually react with the enzyme and change it chemically. These inhibitors modify key amino acid residues needed for enzymatic activity. In contrast, reversible inhibitors bind non-covalently and different types of inhibition are produced depending on whether these inhibitors bind the enzyme, the enzyme-substrate complex, or both.

Enzyme inhibitors also occur naturally and are involved in the regulation of metabolism. For example, enzymes in a metabolic pathway can be inhibited by downstream products. This type of negative feedback slows the flux through a pathway when the products begin to build up and is an important way to maintain homeostasis in a cell. Other cellular enzyme inhibitors are proteins that specifically bind to and inhibit an enzyme target. This can help control enzymes that may be damaging to a cell, such as proteases or nucleases. A well characterized example is the ribonuclease inhibitor, which binds to ribonucleases in one of the tightest known protein–protein interactions. Natural enzyme inhibitors can also be poisons and are used as defenses against predators or as ways of killing prey.

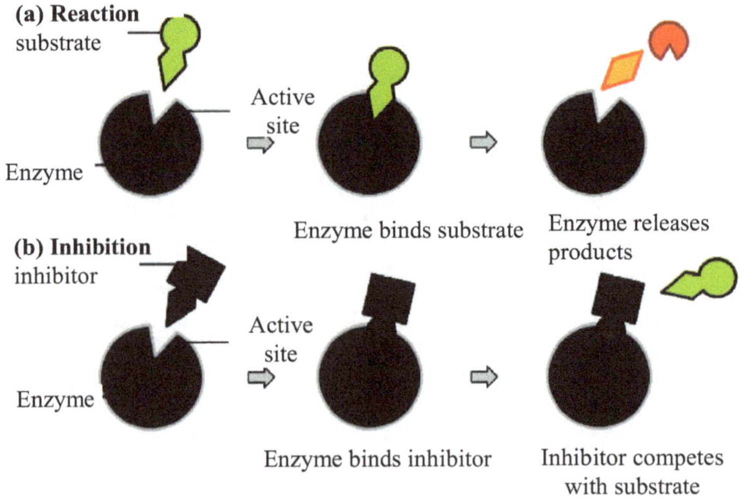

a) **Competitive inhibitor:** Any compound which possesses a close structural resemblance to a particular substrate and which competes with that substrate for the same active site on the enzyme is called a competitive inhibitor. The inhibitor is not acted upon by the enzyme and so remains bound to the enzyme preventing the substrate to bind. This is a reversible process. It depends upon the relative concentration of substrate and inhibitor. Competitive inhibition can be completely reversed by addition of large excess of substrate.

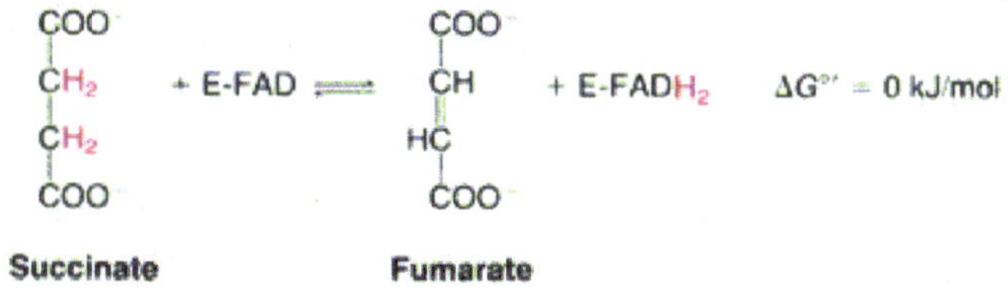

The enzyme, succinate dehydrogenase converts succinic acid to fumaric acid. For this reaction, malonic acid is a competitive inhibitor as it structurally resembles that of succinate.

Glyphosate is a herbicide which resembles phospho enol pyruvate. Glyphosate inhibits EPSP (enoyl pyruvyl shikimate 3 phosphate) synthase involved in shikimate pathway. If shikimate production is affected, lignin synthesis is affected and hence weed is killed.

b) **Non-competitive inhibitor:** Non-competitive inhibitors (I) bind to a site other than the active site on the enzyme often to deform the enzyme, so that it does not form the ES complex at its normal rate. Once formed, the ES complex does not decompose at the normal rate to yield products. These effects are not reversed by increasing the substrate concentration.

$$E + I \longrightarrow EI$$

$$ES + i \longrightarrow ESI$$

Some enzymes possessing an essential -SH group are noncompetitively inhibited by heavy metal ions (Hg^{++}, Pb^{++}). Some metallo-enzymes are inhibited noncompetitively by metal chelating agents like ethylene diamine tetraacetic acid (EDTA).

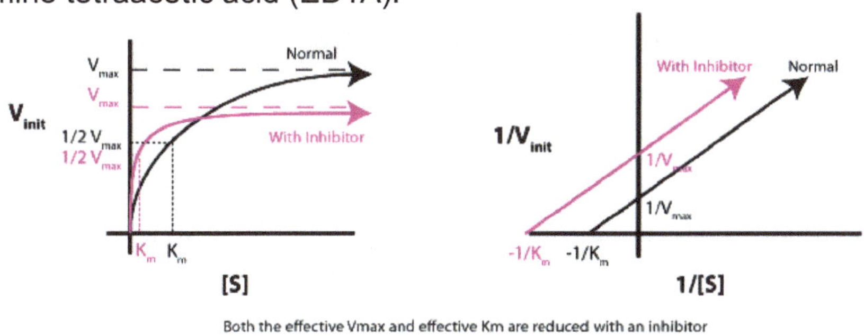

Both the effective Vmax and effective Km are reduced with an inhibitor

c) **Uncompetitive inhibitors :** In this type of inhibition the inhibitor binds to the enzyme substrate complex only and not to the free enzyme. It distorts the active site. The dissociation of ES to form product is affected. H3P04 is uncompetitive inhibitor to the enzyme succinyl CoA synthetase which acts on the substrate succinyl CoA. Both Vmax and Km are decreased. Inhibitors are used as tools to probe the mechanism of enzyme -catalyzed reactions and as therapeutic agents.

Some poisons: Cyanide is a deadly poison. It acts instantly, even at very micro levels. The fatal dose is 50-60mg. Cyanide blocks the oxidation of glucose inside a cell by forming a stable complex with the enzymes. The enzymes cytochrome oxidases contain iron and copper. Cyanide brings about an abrupt end to respiration process by electrons for reduction of oxygen. Hypo is an antidote which must be given very quickly to block the action of cyanide.

Lead is another metal poison. It reacts with sulfhydroxyl groups of enzymes rendering them inactive. Fortunately, lead gets excreted from our body at 2mg per day. Our intake from air, water and food (milk) is less. Mercury and arsenic are also metal poisons which render enzymes inactive. They can be treated with the calcium salt of EDTA.

Ozone is another compound which renders enzymes inactive. Cysteine is oxidized to cysteic acid. Methionine is oxidized to methionine sufoxide. Tryptophan, without sulfur undergoes ring opening.

Isoenzymes

Enzymes which exist in multiple forms within a single species of organism or even in a single cell are called isoenzymes or isozymes. Such multiple forms can be detected and separated by gel electrophoresis of cell extracts. Since they are coded by genes, they differ in amino acid composition and thus in their isoelectric pH values. Lactate dehydrogenase (LDH) is an example for the isoenzymes which occur in five different forms in the tissues of the human and other vertebrate species. All the five isozymes catalyze the same reaction.

$$Lactate + NAD^{+} = Pyruvate + NADH + H^{+}$$

They have the molecular weight of about 1,34,000 and contain four polypeptides. The five isozymes consist of five different combinations of two different kinds of polypeptides M and H. Kinetic study of lactate dehydrogenase isozymes has revealed that although they catalyze the same reaction, they differ significantly in their Km values for their substrates as well as Vmax values. The two polypeptide chains in LDH are coded by two different genes. Skeletal muscle contains four identical M chains and designated as M4, whereas heart muscle contains four identical H chains and designated as H4. LDH of other tissues are a mixture of the five possible forms H4, H3M, H2M2, HM3 and M4. Isoenzymes can be separated by electrophoresis.

Multi-enzyme complexes

A group of related enzymes participating in a given metabolic pathway is called as multi-enzyme complex. This system catalyses sequentially consecutive reactions. The reactions are linked by common intermediates so that the product of first enzyme is substrate for the second enzyme. Ex: All the enzymes catalyzing glycolysis will form into a complex called metabol. The enzymes catalyzing synthesis of fatty acid will form a complex of seven proteins namely, Fatty Acid Synthetase complex

Allosteric enzymes

These enzymes are specialized to serve a regulatory function in addition to their catalytic activity. They have distinct regulatory and catalytic sites. In many multienzyme systems, the end product of the reaction sequence may act as a specific inhibitor of an enzyme at or near the beginning of the sequence. This type of inhibition is called end product inhibition or feed back inhibition or retro-inhibition. The first enzyme in this sequence that is inhibited by the end product is called an allosteric enzyme. The term allosteric denotes another space or another structure. Ex: L-Threonine dehydratase, the first enzyme in the synthesis of L-isoleucine is inhibited by the L-isoleucine. This inhibition can adjust continually the rate of synthesis of metabolic intermediates. Another interesting feature of allosteric enzymes is that they usually do not show classical Michaelis – Menten kinetic relationships between substrate concentration, Vmax and Km because their kinetic behavior is greatly altered by variations in the concentration of the allosteric modulator. On the other hand some allosteric enzymes show sigmoid curves.

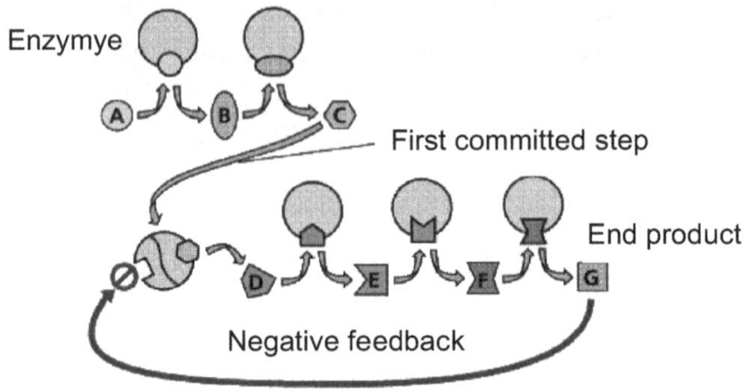

Coenzymes

The non protein component of the enzyme is called as coenzyme. They are loosely bound to the enzyme or non covalently bound. Whatever changes occur on the substrate the opposite changes occur on the coenzyme. Hence coenzymes are also called as cosubstrates. Ex: Glucose is converted to glucose 6 phosphate which is catalysed by hexokinase in the presence of coenzyme ATP. Here glucose is gaining a phosphate molecule and ATP is loosing a phosphate and thereby forming ADP.

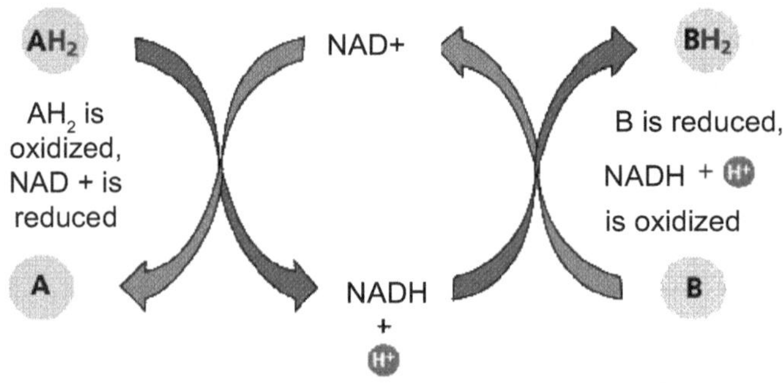

Vitamins: Some of the vitamins act as coenzymes. Vitamins are organic molecules that are essential for the normal health and growth. They are required in trace amounts and must be obtained from the diet as they are not synthesized in the body. Water soluble vitamins are required by many enzymes as cofactors to carry out their catalytic activity. The coenzymes do not remain bound to a particular enzyme. They are used repeatedly by the enzyme. Hence they are required in small quantities in cells.

Table 1. Vitamins and their coenzymes with their metabolic roles

Vitamins	Coenzymes	Function
Thiamine(B_1)	Thiamine pyro phosphate	Decarboxylation
Riboflavin(B_2)	FAD, FMN	Electron transfer
Niacin(B_3)	NAD, $NADP^+$	Oxidation, reduction
Pantothenic acid(B_5)	Coenzyme A	Acetyl group transfer
Pyridoxine(B_6)	Pyridoxal phosphate	Transamination
Cobalamine(B_{12})	Methyl cobalamine	Methyl group transfer
Ascorbic acid(C)	Vitamin C	Collagen synthesis, wound healing
Biotin	Biocytin	Carboxylation
Folic acid	Tetra hydrofolate	Methyl group transfer
Lipoic acid	Lipoic acid	Hydrogen atom transfer & acyl group transfer

Classification of enzymes

The International Union of Biochemistry (IUB) established a commission on enzyme nomenclature to adopt a systematic classification and nomenclature of all the existing and yet to be discovered enzymes. This system is based on the substrate and reaction specificity. Although, this International Union of Biochemistry (IUB) system is complex, it is precise, descriptive and informative. IUB system classifies enzymes into six major classes: 1. Oxidoreductases, 2. Transferases, 3. Hydrolases, 4. Lyases, 5. Isomerases and 6. Ligases. Again each class is divided into subclasses according to the type of reaction catalysed. Each enzyme is assigned an enzyme commission number(E.C), a recommended name usually a short cut for day to day use, a systematic name which identifies the reaction it catalyses and a classification number which is used where accurate and unambiguous identification of an enzyme is required.

1.**Oxidoreductases:** These are the enzymes which catalyse oxidation-reduction reactions between two substrates. Oxidation is the reverse of reduction and an enzyme catalyzes the forward and reverse reactions to the same degree. Any enzyme which catalyzes a reduction has to also catalyze the reverse (oxidation) reaction. Thus the double-barreled name "oxidoreductase" is regularly used. Example: Alcohol dehydrogenase enzyme catalyses the conversion of ethyl alcohol to acetaldehyde.

$$CH3\ CH2\text{-}OH + NAD+ \longrightarrow CH3CHO + NADH + H+$$

ETHANOL COENZYME ACETALDEHYDE +COENZYME

Enzyme: Recommended name Alcohol Dehydrogenase

Systematic name Alcohol:NAD+ oxido-reductase

Enzyme Commission number E.C.1.1.1.1

First digit 1 indicates oxido-reductase (Major class)

Second digit 1 indicates enzymes acting on CH-OH group of donors (Sub-class)

Third digit 1 indicates NAD+ as the electron acceptor (Sub-sub- class),

Fourth digit 1 indicates the specific enzyme

2. **Transferases:** These are the enzymes which catalyze the transfer of a functional group other than hydrogen between substrates. Examples are transaminases and kinases. Phosphorylation of glucose by hexokinase and transfer of phosphate between ATP and sugar molecule. Another example is the transamination of alpha ketoglutarate from alanine to form pyruvate and glutamate by alanine transaminase.

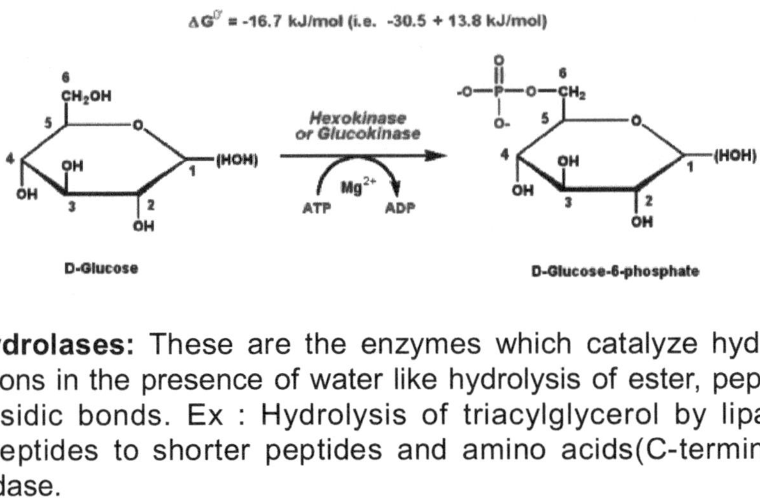

3. **Hydrolases:** These are the enzymes which catalyze hydrolysis reactions in the presence of water like hydrolysis of ester, peptide or glycosidic bonds. Ex : Hydrolysis of triacylglycerol by lipase or polypeptides to shorter peptides and amino acids(C-terminal) by peptidase.

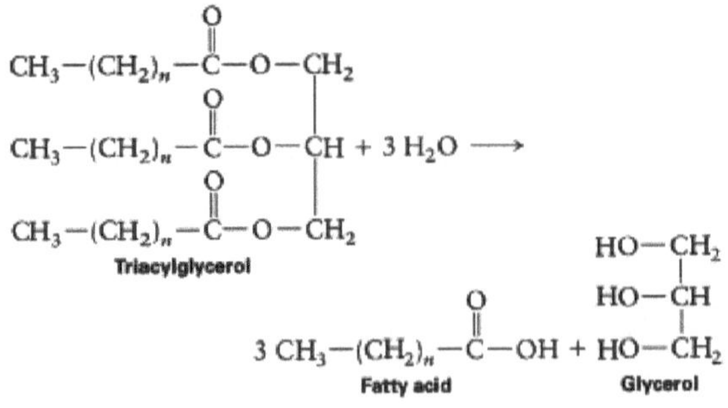

4. **Lyases:** These are the enzymes which catalyze the removal of groups from substrates by a mechanism other than hydrolysis leaving a double bond in one of the products. Reactions which add a small molecule such as water or ammonia to a double bond (and the reverse, elimination, reactions) are also catalyzed by lyases. Example: Aldolase, decarboxylase, etc. Pyruvate in an acid medium is converted to acetaldehyde and carbon dioxide by pyruvate decaboxylase. Decarboxylases remove CO_2, dehydrases remove H_2O, while deaminases remove NH_3.

$$\underset{\text{pyruvate}}{CH3-\overset{\overset{O}{\|}}{C}-COO^- + H^+} \xrightarrow{\text{Pyruvate decarboxylase}} \underset{\text{Acetaldehyde}}{CH3-\overset{\overset{O}{\|}}{C}-H + CO2}$$

5. **Isomerases:** These are the enzymes which catalyze the inter conversion of optical, geometrical or positional isomers. The cis-trans interconversion of maleate and fumarate is an example. The conversion of glucose-6-phosphate to frucrose-6-phosphate is another example. Rearrangement of atoms is carried out to form isomers. Conversion of D- isomer to L- isomer is another function of this enzyme.

$$\text{GLUCOSE-6-PHOSPHATE} \underset{\text{ISOMERASE}}{\overset{\text{PHOSPHO GLUCOSE}}{\rightleftharpoons}} \text{FRUCTOSE-6-PHOSPHATE}$$

6. **Ligases:** These are the enzymes which catalyze the joining together (ligate) of two compounds with the hydrolysis of a high energy compound. Small molecules are ligated to make bigger ones. Synthetases and carboxylases are the main enzymes of ligases. An example is the conversion of pyruvate and CO_2 in the presence of ATP to form oxaloacetate.

$$\textbf{Pyruvate + CO2 +ATP} \longrightarrow \textbf{Oxaloactate +ADP +Pi + H+}$$

Another example is the conversion of glutamate to glutamine by glutamine synthetase.

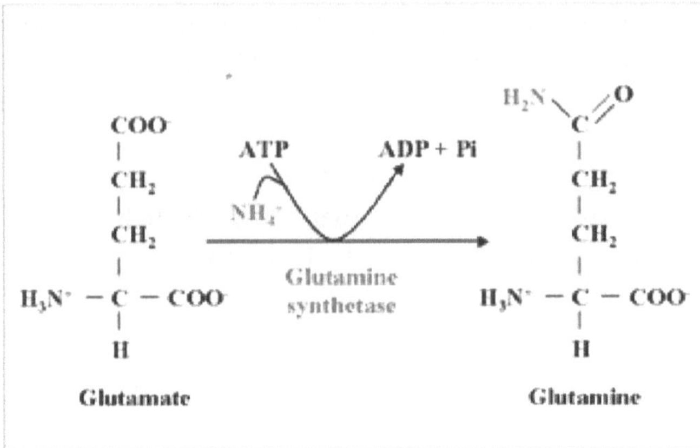

Regulation of enzyme activity

In enzyme catalyzed reactions, the quantity of products produced is based on the need of the cell and never in excess. This is because the enzymes when they are produced are in an inactive state and are called zymogens. These are then transported and activated at the site of need. Zymogens contain longer protein chains and are activated by removal of peptide sections. For example, an enzyme called enteropeptidase removes a hexapeptide from trypsinogen(digestive enzyme) to give acive trypsin. Trypsin in turn cleaves peptide sections from chymotrypsinogen to form active chymotrypsin. Most protein hormones like insulin, enzymes involved in digestion, and enzymes that catalyze blood clotting are initially synthesized as zymogens.

$$\text{CHYMOTRYPSINOGEN} \xrightarrow{\text{TRYPSIN}} \text{CHYMOTRYPSIN + 2DIPEPTIDES}$$

Immobilization of enzymes and their uses

Immobilized enzymes are enzymes which may be attached to each other, or to insoluble materials, or enclosed in a membrane or gel. This can provide increased resistance to changes in conditions such as varying pH or temperature. It also allows enzymes to be held in place throughout the reaction, following which they are easily separated from the products and may be used again.

Advantages of immobilisation: 1. Easier purification of the product as the separation of the enzymes from the products is easily accomplished. 2. It is easy to recover and recycle the enzymes. This leads to a more economical process. 3. The enzymes remain functional for much longer time as it is a gentler process. 4. Immobilized enzymes are used in bioreactors.

Methods of Immobilization: When immobilizing an enzyme to a surface, it is most important to choose a method of attachment that will prevent loss of enzyme activity. The surface on which the enzyme is immobilized is responsible for retaining the structure in the enzyme through hydrogen bonding or the formation of electron transition complexes. The following methods of immobilization of enzymes are generally used. a) Carrier-Binding b) Cross-Linking c) Entrapping and d) Adsorption

a) **Carrier-Binding:** In this method, the amount of enzyme bound to the carrier and the activity after immobilization depend on the nature of the carrier. The selection of the carrier depends on the nature of the enzyme itself, as well as the particle size, surface area, molar ratio of hydrophilic to hydrophobic groups and chemical composition. Some of the most commonly used carriers for enzyme immobilization are polysaccharide derivatives such as cellulose, dextran, agarose, and polyacrylamide gel.

b) **Cross-Linking:** Immobilization of enzymes has been achieved by intermolecular cross-linking of the protein, either to other protein molecules or to functional groups on an insoluble support matrix. The most common reagent used for cross-linking is glutaraldehyde.

c) **Entrapping:** The entrapment method of immobilization is based on the localization of an enzyme within the lattice of a polymer matrix or membrane. It is done in such a way as to retain protein while allowing penetration of substrate.

d) **Adsorption:** In this method the enzyme is attached to a support. Supports can be ceramics, glass, or plastics.

Industrial and other application of enzymes

1. **Starch hydrolysis:** The traditional acid hydrolysis of starch has been completely replaced by α -amylases and glucoamylases, which convert starch with over 95% yield to glucose. α-amylase cuts rapidly the large alpha- 1, 4-linked glucose polymers into shorter oligomers at high temperatures. Glucoamylase hydrolyses the oligomers into glucose. Sometimes additional debranching enzymes like pullulanase are added to improve the glucose yield.

β-amylase is commercially produced from barley grains and used for the production of the disaccharide maltose.

2. **Baking:** α -amylases have been most widely studied in connection with improved bread quality and increased shelf life. In addition to starch, flour typically contains minor amounts of cellulose, glucans and hemicelluloses like arabinoxylan and α arabinogalactan. The use of xylanases decreases the water absorption and thus reduces the amount of added water needed in baking. This leads to a more stable dough. Proteinases can be added to improve dough handling properties. Glucose oxidase has been used to replace chemical oxidants and lipases to strengthen gluten, which leads to more stable dough and better bread quality.

3. **Detergents:** Detergents have been the first large scale application for microbial enzymes. Bacterial proteinases are still the most important detergent enzymes. Lipid degrading enzymes are being added in powder and liquid detergents. Lipases decompose fats into more water soluble compounds by hydrolyzing the ester bonds between the glycerol backbone and fatty acid. Amylases are used in detergents to remove starch based stains. In textile washing, cellulases remove cellulose microfibrils.

4. **Pulp and Paper:** The major application is the use of xylanases in pulp bleaching. Xylanases liberate lignin fragments by hydrolyzing residual xylan. This reduces considerably the need for chlorine based bleaching chemicals. Paper making enzymes are used especially in modification of starch, which is used as an important additive. Starch improves the strength, stiffness and erasability of paper. The starch suspension must have a certain viscosity, which is achieved by adding amylase enzymes in a controlled process.

5. **Leather:** Leather industry uses proteolytic and lipolytic enzymes in leather processing. The use of these enzymes is associated with the structure of animal skin as a raw material. Enzymes are used to remove unwanted parts. Alkaline proteases are added in the soaking phase. This improves water uptake by the dry skins, removal and degradation of protein, dirt and fats and reduces the processing time. In dehairing and dewooling phases enzymes are used to assist the alkaline chemical process. This results in a more environment friendly process and improves the quality of the leather (cleaner and stronger surface, softer leather, less spots). The used enzymes are typically alkaline bacterial proteases.

6. **Animal feed industry:** Enzymes are used in the preparation of poultry feed, pig feed and turkey feed.

7. **Medicine:** Enzymes are also used in medicine. Factor X_a an enzyme present in the blood converts prothrombin to thrombin. This thrombin converts fibrinogen to fibrin, which is nothing but blood clot. Factor X_a is absent in some people, which is hereditary. Russels viper converts prothrombin to thrombin. Hence the venom of this snake is used in the treatment of hemophilia.

$$\text{PROTHROMBIN} \xrightarrow{\text{FACTOR Xa}} \text{THROMBIN}$$

$$\text{FIBRINOGEN} \xrightarrow{\text{THROMBIN}} \text{FIBRIN}$$

8. **Diagnosis:** ELISA is a technique used in analysis of hormones and toxins. Enzymes are added to the sample containing hormone into an antibody coated tube. The free hormone with the enzyme is measured colorimetrically.

9. **Other applications:** Lipases are used in the manufacture of cheese, milk chocolate and by vegetable oil industry. Glucose oxidase is used to remove glucose in food products like beer, cheese, soda, dried eggs, wines, fruit juices etc. to increase the shelf life. Some enzymes are used to prevent browning of foods. Lipoxigenases need to be deactivated to stop production of off flavours.

Chapter – 8

Nucleic Acids

Nucleic acids were first discovered in 1868 by Friedrich Meischer. Nucleic acids are high molecular weight polymers that are present in our cells. They store and transfer genetic material from generation to generation. Knowledge of how genes are expressed and how they can be manipulated is becoming increasingly important for understanding of nearly every aspect of biochemistry. These macromolecules pass on information for cellular growth and reproduction. All the genetic information in the cells is called the genome. Every time a cell divides, the information is copied and passed on to the new cells.

Nucleic acids fall into two main classes according to the type of sugar they contain: the Deoxyribonucleic acids (DNA) and Ribonucleic acids (RNA). Some sections of the DNA called genes contain the information to make a protein. RNA translates the genetic information in the DNA to the ribosomes, where the synthesis of proteins takes place.

Functions of nucleic acids

a) DNA stores and transmits genetic information

b) DNA expresses its encoded genetic information for the synthesis of RNA and protein for metabolic function.

c) DNA controls all cellular activities.

d) RNA is necessary for protein biosynthesis.

Components of nucleic acids

Nucleic acids are polymers of repeating units called nucleotides. Nucleotides are composed of nitrogenous base, sugar and phosphoric acid. Nucleotides in nucleic acids are linked by 3'5'phospho diester linkages.

Nitrogen Bases: The five bases present in nucleic acids have carbon – nitrogen ring structures. They are called nitrogen bases. There are two types of ring structures, purines and pyrimidines. Adenine and Guanine are purine nitrogen bases. Both these bases are present in both DNA and RNA. Thymine, Cytosine and Uracil are pyrimidine nitrogen bases. Pyrimidine nitrogen base present in both RNA and DNA is Cytosine, that present only in RNA is Uracil and present only in DNA is Thymine. A pyrimidine or purine is linked to the sugar ribose or deoxyribose to form a nucleoside. The phosphoric acid esters of nucleosides are called nucleotides. Both DNA and RNA are polynucleotides.

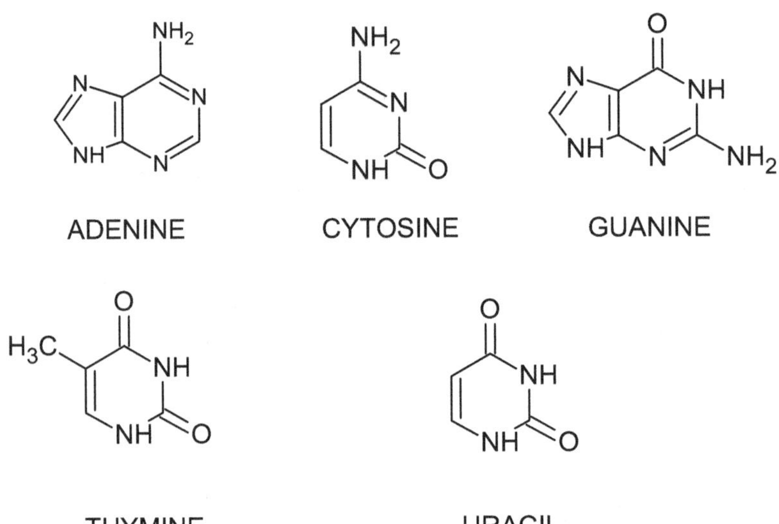

Fig. 1. Nitrogen bases of nucleic acids

Nucleosides: The nitrogen bases are combined with ribose/deoxyribose to form nucleosides. In RNA, the nucleosides have ribose as sugar component and are called ribonucleosides, whereas they are called deoxyribonucleosides in DNA as the sugar in DNA is deoxyribose. Nucleoside is formed through a bond between C1 of β sugar and N1 of the pyrimidine base or N9 of the purine base. This linkage is called as β - N glycosidic linkage. Adenine and ribose combine to form a nucleoside called adenosine, while cytocine and ribose combine to form cytidine.

Sugar + Base = Nucleoside

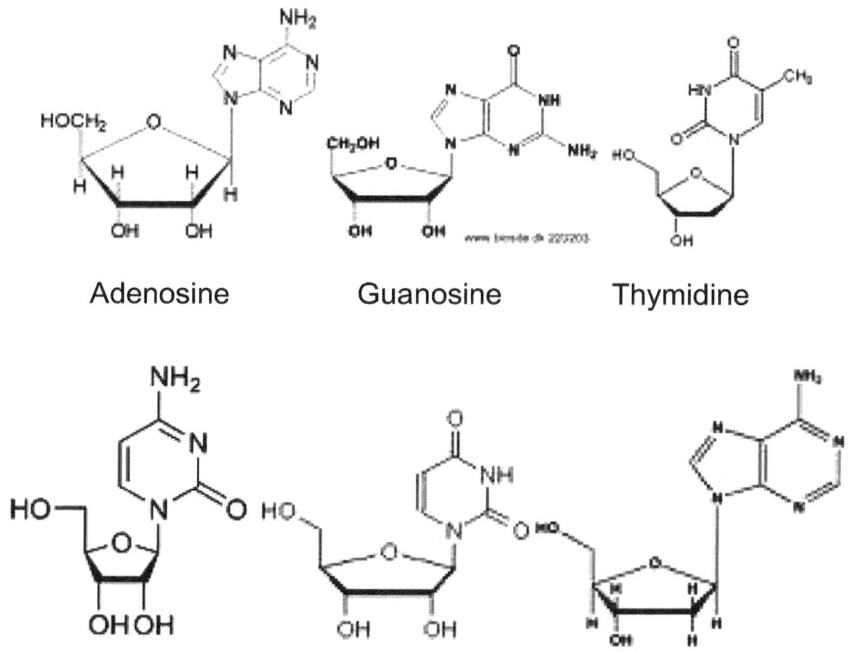

Adenosine Guanosine Thymidine

Cytidine *Uridine* *Deoxyadenosine*

Fig. 2. Nucleosides

Nucleotide: A nucleotide is a phosphate ester of nucleoside. It consists of a phosphate group joined to a nucleoside at hydroxyl group attached to the C5 group of the sugar that is 5'-nucleotide. In DNA, the nucleotides have deoxyribose as the sugar and hence are called deoxyribo-nucleotides. In RNA, the nucleotides have ribose as sugar moiety and hence are called ribonucleotides. Deoxyribonucleotides and ribonucleotides can have a single phosphate group, two phosphate groups or three phosphate groups. Nucleotides are acids. They are negatively charged at a neutral p^H. The di and tri phosphate esters of adenosine 5' –nucleotides are referred to as adenosine diphosphate (ADA) and adenosine triphosphate (ATP).

Nucleoside + Phosphoric acd = Nucleotide (phosphate ester of nucleoside)

Functions of nucleotides

1. They play a major role in energy metabolism. Ex. ATP
2. They are the building blocks of nucleic acids.
3. They play a role as physiological mediators. AMP (Adenosine Mono Phosphate) plays an important role as a secondary messenger
4. The nucleotides also serve as carriers of activated intermediates. Ex. UDP glucose, CDP choline
5. Many of the regulated steps of metabolic pathways are controlled by the intracellular concentration of nucleotides and behave as allosteric effectors.

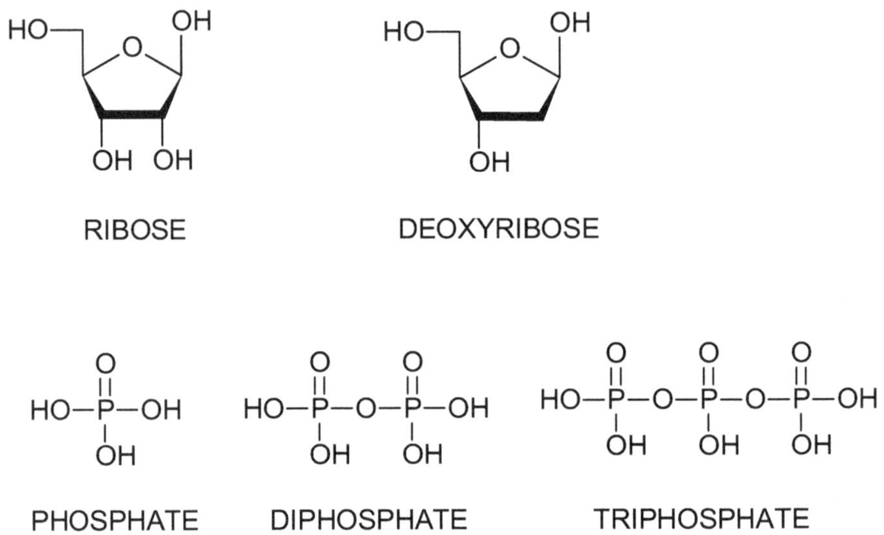

Fig.3. Sugars and phosphates in nucleic acids

Table 1. Hydrolysis products of DNA and RNA

Components	DNA	RNA
Purine bases	Adenine, guanine	Adenine, guanine
Pyrimidine bases	Cytosine, thymine	Cytosine, uracil
Pentose sugar	2-deoxy ribose	Ribose
Inorganic acid	Phosphoric acid	Phosphoric acid

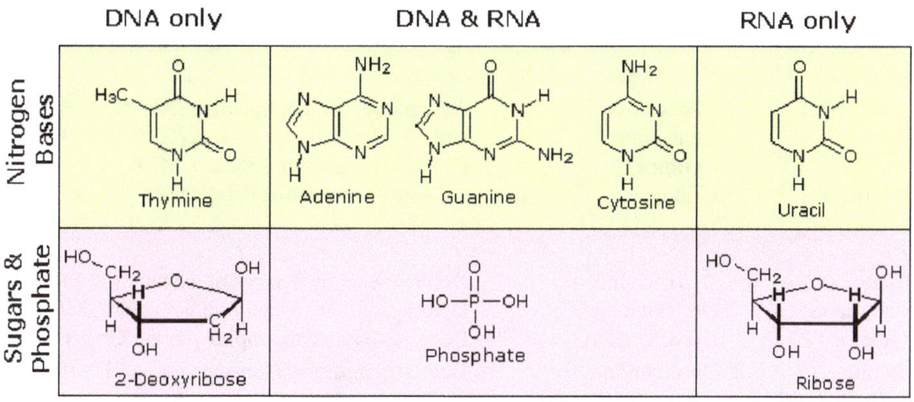

Fig. 4. Components of nucleic acids

Naming of nucleotides: The name of nucleotide is obtained from the name of the nucleoside followed by 5' monophosphate. For example in DNA it is named deoxyadenosine monophosphate or dAMP. In RNA it is called adenosine monophosphate (AMP). The names of the different nucleosides and nucleotides are presented in the Table 2. The letters A,G,C,U and T represent the different bases.

Nucleic acids: In nucleic acids, the different nucleotides are joined to form a long polymer by 3'-5' Phosphodiester bond between phosphates and sugars. Each sugar has at least two hydroxyl groups, the 3' and 5' hydroxyl groups of Ribonucleotides and Deoxyribonucleotides. Each phosphoric acid has three esterifiable acidic groups. The 3'-5' Phosphodiester bond between nucleotides joins only the 3' hydroxyl of one nucleotide to 5' hydroxyl of another nucleotide. Hence each nucleotide in a polynucleotide chain has its 3' hydroxyl linked by a phosphate diester bond to the 5' hydroxyl of another nucleotide and the 3' hydroxyl of this nucleotide is linked to the 5' hydroxyl of still another nucleotide except at the end of the chain. This phosphate link between the sugars in adjacent nucleotides is referred to as a phosphodiester bond. The first nucleotide has a free 5' phosphate and the last nucleotide has a free 3' hydroxyl group

Table 2. Nucleosides and nucleotides in DNA and RNA

Base	Nucleosides	Nucleotides
RNA		
Adenine(A)	Adenosine	Adenosine5'monophosphate(AMP)
Guanine(G)	Guanosine	Guanosine5'monophosphate(GMP)
Cytocine(C)	Cytidine	Cytidine5'monophosphate(CMP)
Uracil(U)	Uridine	Uridine5'monophosphate(UMP)
DNA		
Adenine(A)	Deoxyadenosine	Deoxyadenosine5'monophosphate (dAMP)
Guanine(G)	Deoxyguanosine	Deoxyguanosine5'monophosphate(dGMP)
Cytocine(C)	Deoxycytidine	Deoxycytidine5'monophosphate(dCMP)
Thymine(T)	Deoxythymidine	Deoxythymidine5'monophosphate(dTMP)

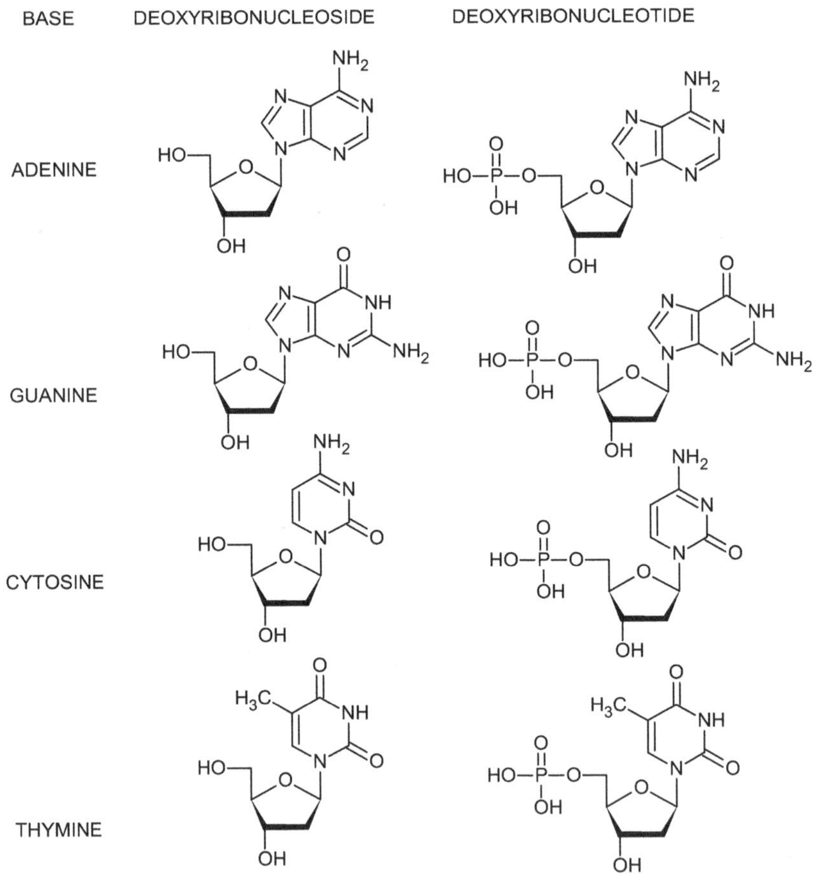

Fig. 5. Structures of nucleosides and nucleotides in DNA

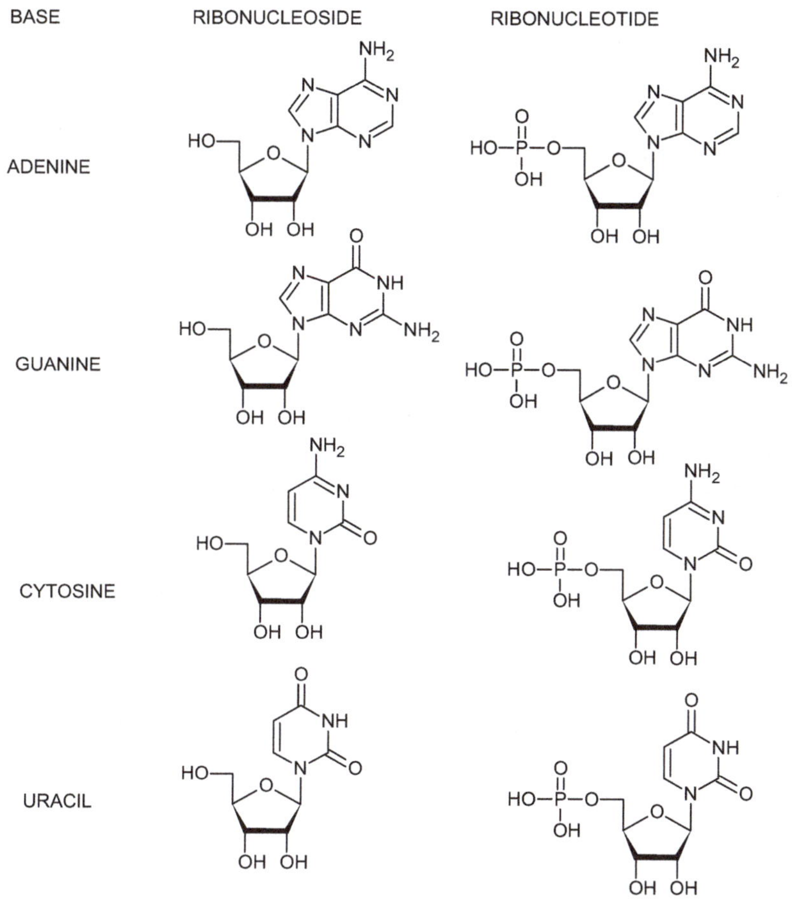

Fig. 6. Strucures of Nucleosides and nucleotides in RNA

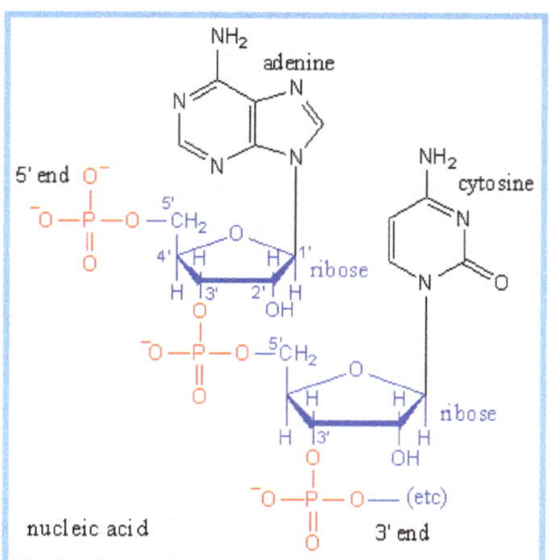

Fig. 7. 3'-5' Phospho diester bond formation

Structure of DNA: In 1953, Watson and Crick worked out the three-dimensional structure of DNA. It is a duplex molecule. DNA is composed of two strands of poly deoxyribonucleotide chains wound around each other to form a double helix, with the bases on the inside and the sugar-phosphate backbones on the outside. In the double helix, the two DNA strands are organized in an anti-parallel arrangement (i.e. the two strands run in opposite directions, one strand is oriented 5' : 3' and the other is oriented 3' : 5').

DNA in a variety of organisms has a specific relationship between bases. The percent of adenine(A) is equal to the percent of thymine(T) and the percent of guanine (G) is equal to cytosine(C). Different organisms vary in the amounts of bases. But the pairing namely A-T and G-C remains the same in all organisms in the ratio of 1:1. This relationship is known as *Chargaff's rule.* The composition of the bases in DNA of some organisms is presented in the Table 3.

Table 3. Per cent bases in the DNAs

Organism	%A	%T	%G	%C
Humans	30	30	20	20
Chicken	28	28	22	22
Salmon	28	28	22	22
Corn(maize)	27	27	23	23
Neurospora	23	23	27	27

The nitrogen bases of the two strands form hydrogen bonds to each other; Adenine base pairs with Thymine and Guanine base pairs with Cytosine. This is called complementary base pairing. Thus a large two-ringed purine is paired with a smaller single-ringed pyrimidine and the two bases fit neatly in the gap between the sugar-phosphate strands and maintain the correct spacing. The G - C and A - T base pairing also maximizes the number of effective hydrogen bonds that can form between the bases. There are three hydrogen bonds between each G – C base pair and two hydrogen bonds between each A - T base pair. Thus A - T and G - C base pairs form the most stable conformation from the point of view of maximizing hydrogen bond formation. The spatial relationship between the strands creates a major groove and a minor groove.

Since each base pair contains a purine and a pyrimidine, the total width of the two pairs of bases A-T, and G-C is the same. Hence the two polynucleotide strands of DNA are at the same distance all through the

DNA polymer. There are no C-T, C-A, G-T, or G-A pairs in DNA since the combinations cannot form many hydrogen bonds or maintain a constant width between the two strands of DNA. X-ray diffraction studies have indicated that DNA is a right handed or alpha helix. In one turn there are 10 pairs of nucleotides. In mitochondria, bacteria and viruses, DNA molecules are compact and highly coiled molecules. In chromosomes, DNA strands are wrapped around proteins called histones. Such an arrangement provides the most stable and orderly structure to the DNA molecules.

Forms of DNA: Different forms of DNA exist. B form is the most commonly occurring one in nature. In the B form the base pairs are stacked on top of one another, with the plane of the base pairs being perpendicular to the length of the duplex. The double helical structure of the DNA takes a right-handed form called the B-helix. There is one complete twist for every 10 base pairs or 3.4nm. This distance is referred to as the pitch. The surface of the helix shows alternating major and minor grooves. The complimentary strands are oriented in opposite directions, called anti-parallel directions. B-DNA has two principal grooves, a wide major groove and a narrow minor groove. Many proteins interact in the space of the major groove, where they make sequence-specific contacts with the bases. In addition, a few proteins are known to make contacts via the minor groove.

If some of the water of hydration in the B form is removed, the double helix becomes more closely packed (11 base pairs instead of 10) which is called A form. It has shorter and wider helix than the B- form. The deep and narrow groves are not easily accessible to proteins. It is a right handed helix with a size of 11 A^0.

Another form also exists with 12 base pairs per turn. This Z DNA is left handed helix. The sugar phosphate back bone takes a zig-zag course and hence called Z -DNA. The Z form is about 18 A^0 and there are 12 base pairs per helical turn, and the structure appears more slender and elongated. Z-DNA formation occurs during transcription of genes

Table 4. Forms of DNA and their properties

Character	A-DNA	B-DNA	Z-DNA
Helical	Right handed	Right handed	Left handed
No of base pairs	11	10	12
Helix pitch	2.8nm	3.4nm	4.5 nm
Base tilted to normal axis	20 A^0	6 A^0	7 A^0
Helix rise per base pair	2.4 A^0	3.4 A^0	3.7 A^0
Occurrence	Sprouting bacteria	Common	E.Coli

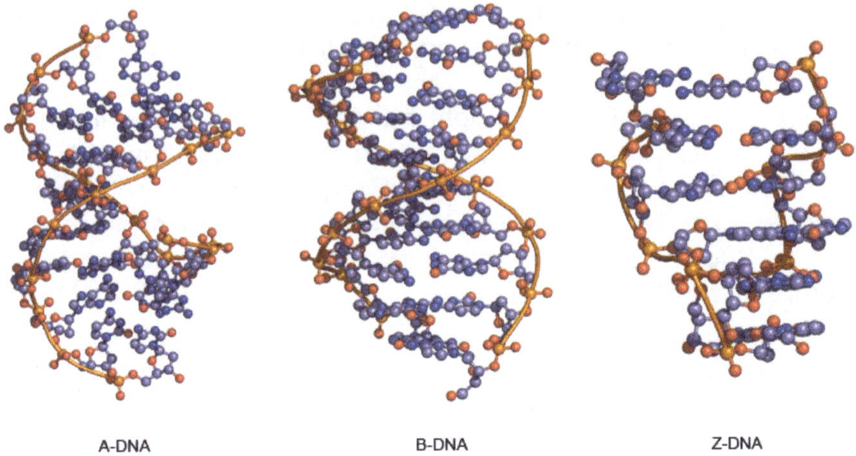

Fig. 8. Different forms of DNA

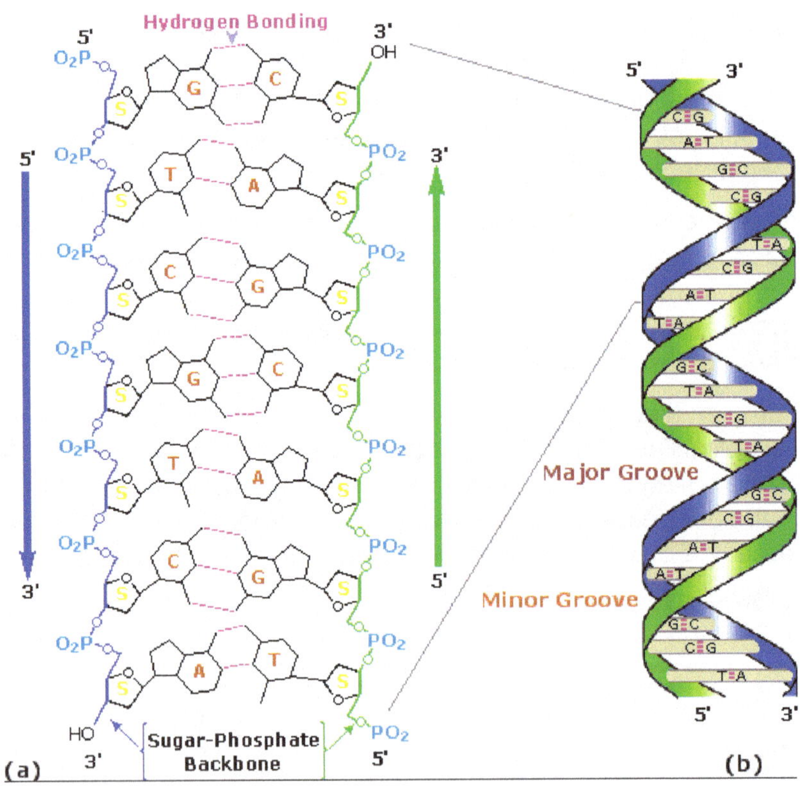

Fig. 9. Double helical structure of DNA

DNA Replication: In DNA replication the strands in the parent DNA separate, followed by synthesis of complementary strands. Within the nucleus, the nucleoside triphosphates for each base are available so that each exposed base on the template strand can form hydrogen bonds with its complementary base in the nucleoside triphosphate. DNA polymerase catalyzes the formation of orthophosphodiester bonds between the nucleotides.

Eventually the entire double helix of the parent DNA is copied. In the new DNA molecule one is the original while the other one is synthesized and added. In the process of DNA replication, complementary base pairing ensures the correct placement of bases in the new DNA strands.

Structure of RNA

Chemically, RNA is very similar to DNA but there are some important differences. RNA consists of the sugar ribose instead of deoxyribose and nitrogen base uracil in place of thymine. Most RNA is single stranded. There are three distinct types of RNAs: Messenger RNA, Transfer RNA and Ribosomal RNA. These differ from one another in size and function. The synthesis of these is catalyzed by enzymes called RNA Polymerases.

Messenger RNA (mRNA): mRNA is the copy of the information carried by a gene on the DNA. The role of mRNA is to carry the information present in the DNA molecule to the translation machinery. This type of RNA makes up to 2% of the total RNA. It consists of the sequences of bases needed to determine the order in which the amino acids will be assembled in to each type of protein. There are several thousand types of mRNA in most cells. In eukaryotes, one type of mRNA corresponds to each type of protein but in prokaryotes, some mRNAs are polycystronic. In this case, an mRNA molecule may carry the code for several different proteins. In prokaryotic cells, mRNA is made within the nucleoid region of cytoplasm and is then used immediately to code for protein synthesis. In eukaryotes, most of the mRNA is made in the nucleus, from where it passes out through the nuclear envelope to the cytoplasm.

Ribosomal RNA (r RNA): Ribosomal RNA makes up to more than 80% of the total RNA of the cell. It is the major component of the ribosomes, the organelles on which protein synthesis takes place. Each ribosome consists of two unequal subunits, a small subunit and a large subunit each of which is a multi component complex of ribosomal RNAs and ribosomal proteins. A prokaryotic ribosome has a sedimentation coefficient of 70 S whereas the large and small subunits have sedimentation coefficients of 50 S and 30 S, respectively. In eukaryotes, the ribosomes are larger and more complex; the ribosome monomer is 80 S and consists of 60 S and 40 S subunits.

Transfer RNA (t RNA): tRNA is the information adaptor molecule. It is the direct interface between amino acid sequence of a protein and the information on DNA. There are more than 20 different t RNA molecules. Transfer RNA makes up about 16% of the total RNA. They are small, uniform in size having 75-90 nucleotides. These molecules play an important role in protein biosynthesis. Each tRNA becomes covalently bonded to a specific amino acid to form amino-acyl - tRNA which recognizes the corresponding codon in mRNA and ensures that the correct amino acid is added to the growing polypeptide chain. The tRNAs are small molecules, with a distinctive clover leaf secondary structures by internal pairing. The stem-loops of the cloverleaf are known as arms. There are 4 arms and 3 loops in t RNA molecule.

- The anti-codon arm contains in its loop the three nucleotides of the anti-codon which will form base-pairs with the complementary codon in mRNA during translation;
- The D or DHU arm (with its D loop) contains dihydrouracil, an unusual pyrimidine;
- Some tRNAs also have a variable arm (optional arm) which is 3-21 nt size.

The other notable feature is the amino acid acceptor stem. This is where the amino acid becomes attached, at the 3' OH group of the 3'-CCA sequence.

The three dimensional structure of tRNAs *is* even more complex (Figure 10) because of additional interactions between the various units of secondary structure.

- The T or TΨC arm (with its T loop) contains another unusual base pseudouracil (denoted Ψ) in the sequence TΨC

Table 5. Types of RNA and their properties

Type	% of total RNA	Function in the cell
Ribosomal RNA	75	Major component of ribosomes
Messenger RNA	5-10	Carries information for protein synthesis from DNA to ribosomes
Transfer RNA	10-15	Brings amino acids to ribosomes for protein synthesis

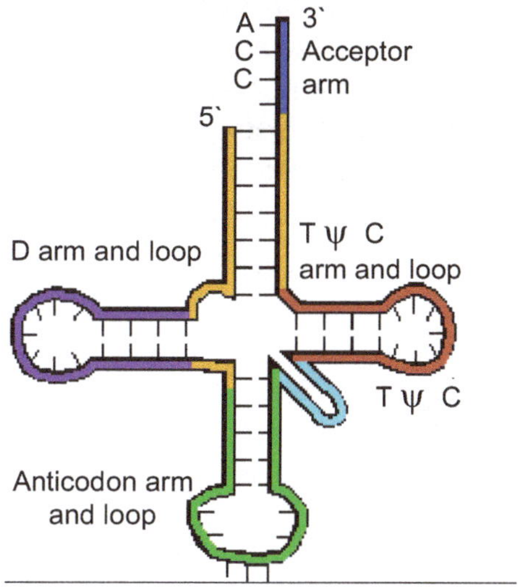

Fig. 10. Structure of tRNA.

Packing of DNA into chromosomes

DNA is a very long molecule and a cell cannot accommodate such long molecules. DNA does not normally exist as the simple double helix. Instead, eukaryotic DNA is packaged in a series of steps with protein, to form chromosomes. The amount of DNA in a single human cell, lined up end to end, would stretch nearly two meters. Hence DNA is compacted enough to fit into a single nucleus. The first level of packaging involves the binding of the chromosomal DNA to histones to from nucleosome.

A nucleosome consists of a small amount of DNA wrapped up with protein. The proteins that interact with DNA to form chromatin comprise a family of basic (positively charged) proteins called histones. There are five different types of histone protein: H1, H2A, H2B, H3, and H4. Of these, two molecules each of H2A, H2B, H3, and H4 combine to form a histone octamer. DNA wraps around the octamer, making 1.8 turns around the protein complex. The amount of DNA associated with the histone octamer is 146 base pairs (bp). The octamer plus the DNA comprise what is called the nucleosome core.

A small stretch of DNA (60 bp) runs between adjacent nucleosome cores, and is known as the linker. A single nucleosome consists of one core plus a linker. The total amount of DNA involved in a single nucleosome is approximately 206 bp. Chromatin therefore consists of DNA wrapped around one histone octamer after another, like a long string of beads. This configuration is known as the 11 nm chromatin fiber. Overall, the packing ratio is about 7 that is the DNA length is shortened about sevenfold by winding around the nucleosome. Complexing with histones protects DNA from the action of Deoxyribonuclease.

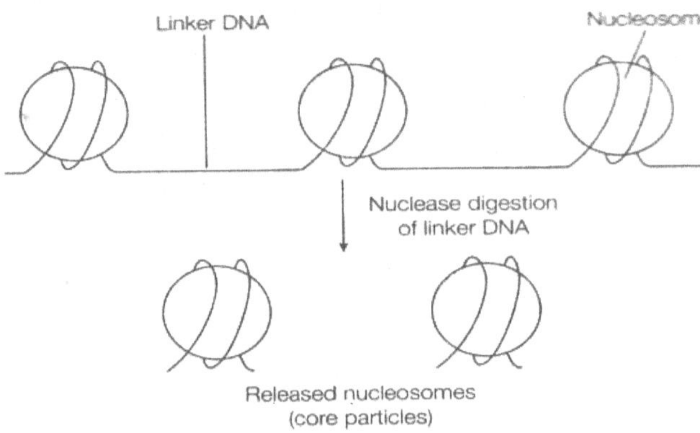

Fig. 11. Structure of chromatin

Chromatin can be packed further into higher-order structures. This involves the action of histone H1. H1 binds to DNA on the outside of nucleosomes (at a ratio of one H1 molecule per nucleosome), then H1 molecules interact with each other, causing the chromatin to form a spiral, with 6 to 8 nucleosomes per turn of the spiral. This structure is known as a solenoid or 30 nm chromatin fiber. Because the chromatin is so tightly packed, DNA in the 30 nm fiber is genetically inactive. The 30 nm fiber is folded up further to make metaphase chromosomes. This 30 nm chromatin fibre is then folded into large looped domain with 300 nm diameter. In the next phase, this is further condensed to form 700 nm diameter looped domains. During mitiotic cycle, the looped domains are organized into distinct structures called the chromosomes with 1400 nm.

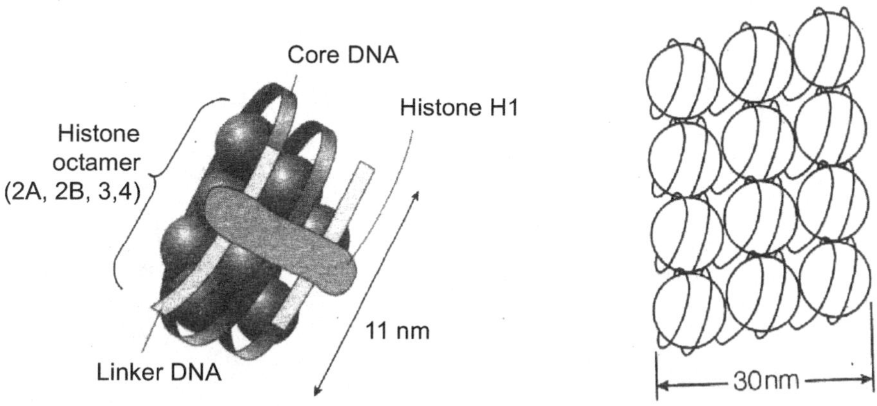

Fig. 12. Nucleosome with DNA and Solenoid structure of Chromatin

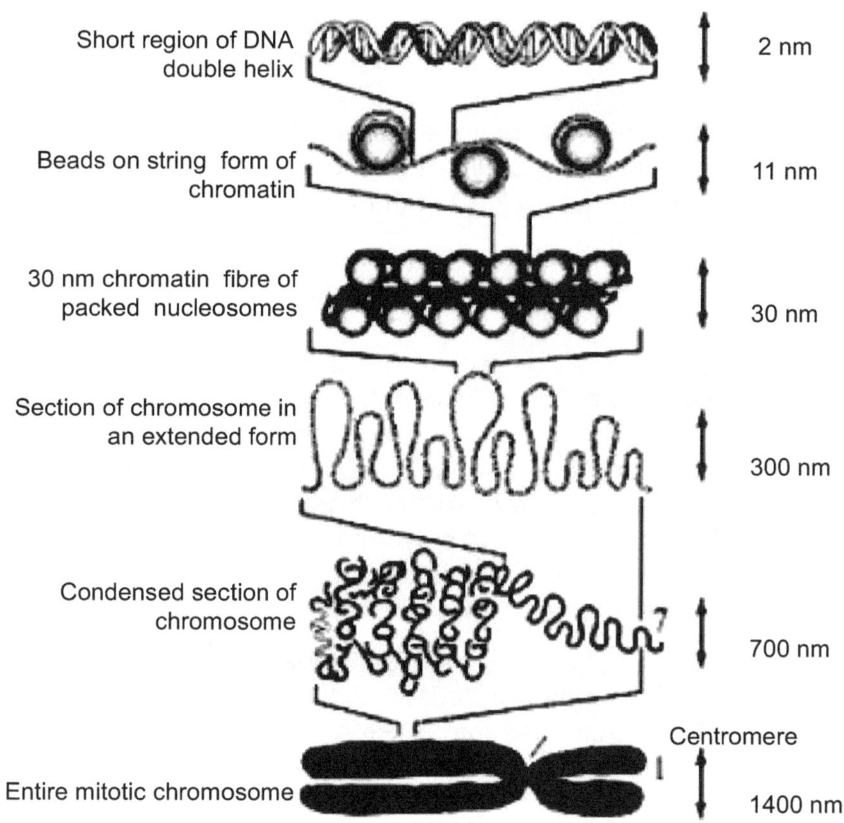

Fig. 13. Packing of DNA in Chromosomes

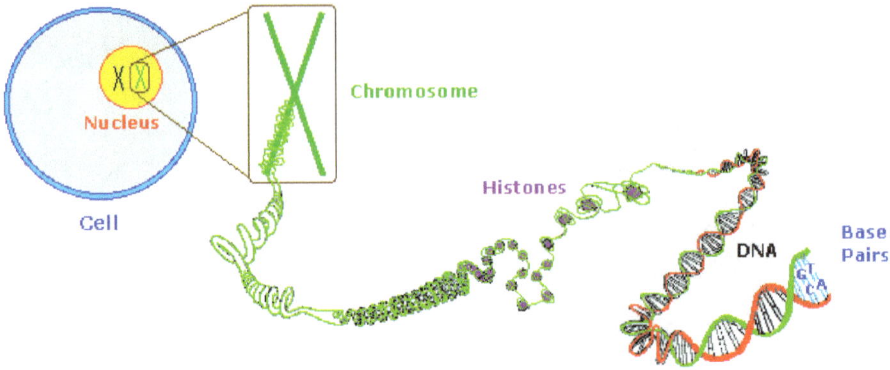

Fig. 14. The Chromosome with histone and DNA

Transcription and protein synthesis: The genetic information is copied from a gene in a DNA to make a messenger RNA, a process called transcription. The m RNA moves out of the nucleus and goes to the ribosomes. During translation the tRNA molecules convert the information in the m RNA into amino acids which are placed in a proper sequence to synthesize a protein. The language that relates to the series of nucleotides in the m RNA with the amino acids specified is the genetic code.

$$DNA \xrightarrow[transcription]{} m\,RNA \xrightarrow[translation]{} Protein$$

Transcription begins when a section of the DNA (gene) unwinds. A polymerase enzyme identifies a nucleotide sequence TATAAAA, which is the initiation and signal point to begin an RNA synthesis. The polymerase moves along the unwound DNA in a 3' to 5' direction forming bonds between the complimentary bases in the nucleoside triphosphates. Eventually, the RNA polymerase reaches the termination point that signals the end of transcription, the new m RNA is released. DNA returns to its double helix structure.

The genetic code: The overall function of the RNAs in the cell is to facilitate the task of synthesizing protein. After the genetic information encoded in DNA is transcribed into m RNA molecules, the m RNA 's move out of the nucleus to the ribosomes in the cytoplasm.

$$m\,RNA \xrightarrow{Translation\ at\ the\ robosomes\ in\ the\ cytoplasm} Protein$$

Codons: Genetic information from DNA is encoded in the m RNA as a sequence of the nucleotides. In the genetic code, a sequence of three bases (triplet) called codon specifies each amino acid in the protein. Repeating triplets of uracil (UUU) produces a polypeptide containing phenylalanine.

Codons in mRNA	**5'- UUU - UUU -UUU- 3'**
Translation	↓ ↓ ↓
Amino acid sequence	**Phe Phe Phe**

The codons for all the 20 amino acids have been determined. A total of 64 codons are possible from the triplet combinations of A,G,C,and U. Three of these, UGA, UAA, and UAG are stop signals that code for the termination of protein synthesis.

Table 6. Codons of m RNA: the genetic code for amino acids.

Firstletter	U	C	A	G	Third letter
U	UUU(Phe)	UCU(Ser)	UAU(Tyr)	UGU(Cys)	U
	UUC(Phe)	UCC(Ser)	UAC(Tyr)	UGC(Cys)	C
	UUA(Leu)	UCA(Ser)	UAA(s)	UGA(s)	A
	UUG(Leu)	UCG(Ser)	UAG(s)	UGG(Try)	G
C	CUU(Leu)	CCU(Pro)	CAU(His)	CGU(Arg)	U
	CUC(Leu)	CCC(Pro)	CAC(His)	CGC (Arg)	C
	CUA (Leu)	CCA(Pro)	CAA(Gln)	CGA (Arg)	A
	CUG(Leu)	CCG(Pro)	CAG(Gln)	CGG(Arg)	G
A	AUU(Iso Leu)	ACU(Thr)	AAU(Asn)	AGU(Ser)	U
	AUC(Iso Leu)	ACC(Thr)	AAC(Asn)	AGC(Ser)	C
	AUA(Iso Leu)	ACA(Thr)	AAA(Lys)	AGA(Arg)	A
	AUG(a)(Met)	ACG(Thr)	AAG(Lys)	AGG(Arg)	G
G	GUU(Val)	GCU(Ala)	GAU(Asp)	GGU(Gly)	U
	GUC(Val)	GCC(Ala)	GAC(Asp)	GGC(Gly)	C
	GUA(Val)	GCA(Ala)	GAA(Glu)	GGA(Gly)	A
	GUG(Val)	GCG(Ala)	GAG(Glu)	GAG(Gly)	G

(a) = start of a peptide chain, (s) = stop codons, end of chain

Chapter – 9

Plant Metabolism

Plants are responsible for highly useful molecular reactions. All the enzymatic reactions that occur in the cell are termed as metabolic reactions. They take place within organisms in an orderly and regulated way. They are the metabolic pathways and the compounds resulting from the process, namely, metabolism are called the metabolites. Metabolism in Greek means change. It is the set of life sustaining chemical transformations within the cells of living organisms. These enzyme-catalyzed reactions allow organisms to grow and reproduce, maintain their structures, and respond to the environments around them.

Metabolism is a highly organized, coordinated and purposeful enzymatic activity inside a cell. Metabolism has four functions: 1. To obtain energy from the environment, 2. To convert molecules into building blocks, 3. To convert the building blocks into proteins, lipids, sugars, nucleic acids and other cell components and 4. To synthesize and degrade organic molecules for specific needs of the cell.

Autotrophs and heterotrophs differ mainly with respect to the initial energy deriving steps. Autotrophs can utilize CO_2 while heterotophs require carbon in the reduced form like sugars etc. Autotrophs can harvest sunlight and convert CO_2 and H2O through photosynthesis to carbohydrates. Heterotrophs depend primarily on organic compounds synthesized by autotrophs.

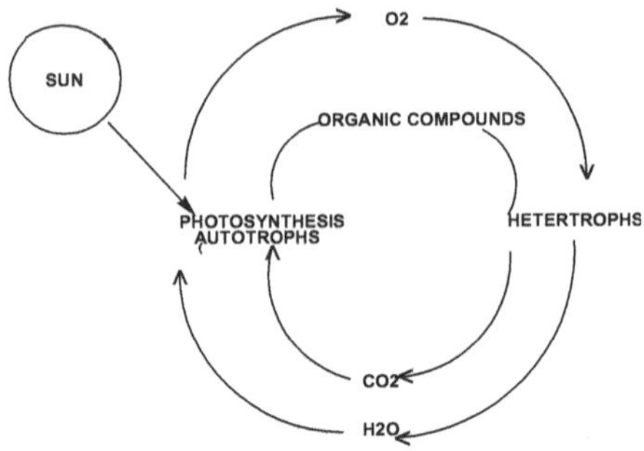

Fig. 1. Autotrophs and heterotrophs.

Reactions comprising the total metabolism of any cell can be broadly divided into two phases, namely, catabolism and anabolism. Catabolism refers to the reactions involving the degradation or breakdown of various complex molecules to smaller and simpler molecules. Catabolism is accompanied by release of free energy present in the large organic molecules. The chemical energy of metabolites is conserved in the form of ATP (adenosine triphosphate). A part of the energy may be conserved in the form of NADPH (nicotinamide adenine dinucleotide-reduced).

Anabolism refers to all the reactions involving synthesis of various molecules from smaller and simpler precursor molecules. This process utilizes chemical energy in the form of ATP generated in the catabolic processes. Large biomolecules like proteins and nucleic acids are built

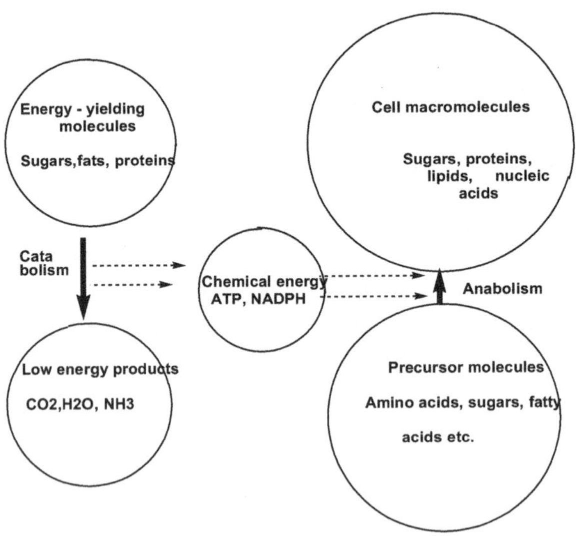

Fig. 2. Overview of metabolism

up during anabolism. NADPH donates the much needed energy for the different steps in the anabolism.

ATP, is called the "energy currency" of the cell. •Energy production and use in living organisms revolve around the ATP - ADP interconversion. Catabolic reactions "pay off" in ATP by synthesizing it from ADP plus

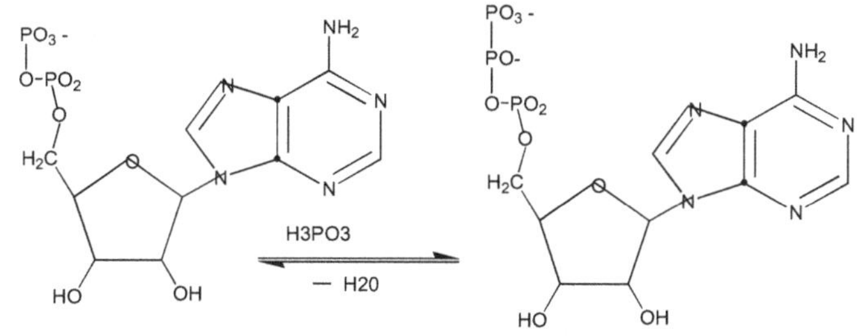

ADENOSINE DIPHOSPHATE(ADP) **ADENOSINE TRIPHOPHATE(ATP)**

phosphate ion. Anabolic reactions "spend" ATP by transferring a phosphate group to another molecule, thereby regenerating ADP.

In a living system, both catabolism and anabolism occur concurrently and simultaneously. Their rates are regulated independently. The energy released during catabolism is required for anabolic reactions and for many other cellular activities. Catabolic reactions are oxidative whereas anabolic reactions are reductive in nature. There is a balance between the two types of reactions to sustain life in an organism.

Catabolism

Catabolism is exergonic and release energy into the system. Catabolic reactions take place in three major stages. Large molecules are broken down to their building blocks in the first stage of metabolism. Polysaccharides are hydrolyzed to yield monosaccharides, lipids are converted to glycerol and fatty acids, and proteins are broken down to amino acids. In the next stage of catabolism, the various products of stage one are collected and converted into a number of more simpler intermediates.

Thus, the monosaccharides and glycerol are degraded via the three carbon intermediate pyruvic acid to yield the two carbon species, the acetyl CoA (co-enzyme A). Similarly, various fatty acids and amino acids are broken down to form acetyl Co A and a few other products. Finally the acetyl Co A and other products are converted/oxidized into carbon

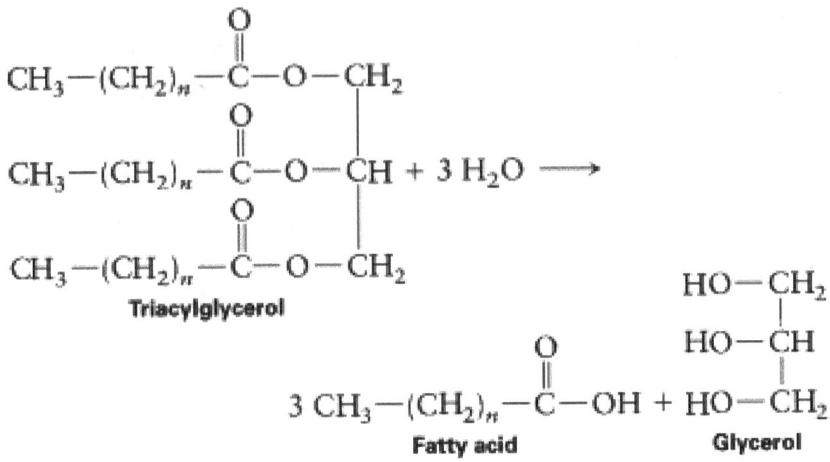

dioxide and water followed by oxidative phosphorylation.

Stage I : Ex: Hydrolysis of fat by lipase.

Stage II: Conversion of monomers into a form that can be completely oxidized.

a) Sugars are converted to puruvic acid and acetyl Co A.

b) Amino acids are deaminated and might enter at any stage into the metabolic pathway.

c) Fatty acids are converted to acetyl Co A

Stage III: Complete oxidation of carbohydrates, fats and proteins and the production of ATP

a) Carbohydrates, fats and proteins are converted to acetyl Co A.

b) The acetyl Co A units are processed in TCA (tricarboxylic acid) cycle and converted to CO2 and water with the liberation of ATP in the oxidative phosphorylation

Overview of catabolic processes

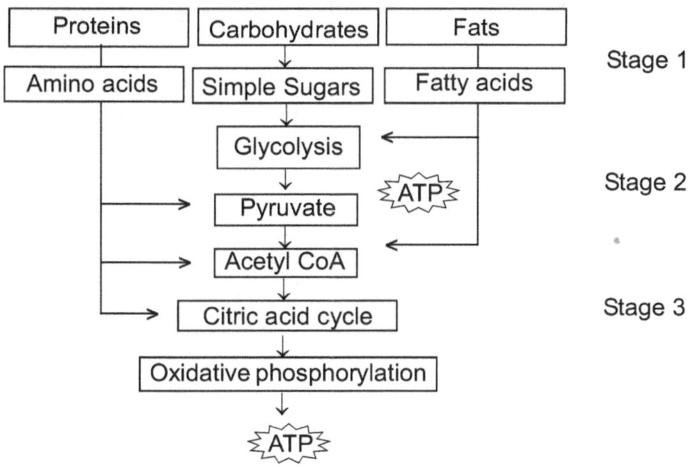

Anabolism

Anabolism is the set of constructive metabolic processes where the energy released by catabolism is used to synthesize complex molecules. Anabolism is endergonic and absorb energy. In general, the complex molecules that make up cellular structures are constructed step-by-step from small and simple precursors. Anabolism involves three basic stages. Firstly, the production of precursors such as amino acids, monosaccharides, isoprenoids and nucleotides, secondly, their activation into reactive forms using energy from ATP, and thirdly, the assembly of these precursors into complex molecules such as proteins, polysaccharides, lipids and nucleic acids.

Organisms differ in the number of molecules they can synthesize in their cells for themselves. Autotrophs such as plants can synthesize the complex organic molecules like polysaccharides and proteins from carbon dioxide and water. Heterotrophs, on the other hand, require a source of more complex substances, such as monosaccharides and amino acids, to produce these complex molecules. Organisms can be further classified by ultimate source of their energy: photoautotrophs and photoheterotrophs obtain energy from light, whereas chemoautotrophs and chemoheterotrophs obtain energy from inorganic oxidation reactions.

Table 1. Differences Between The Two Metabolic Processes:

Anabolism (Photosynthesis)	Catabolism (Respiration)
• Energy is stored with sugar molecules	• Energy is released from sugar molecules
• Water and carbon dioxide are used	• Water and carbon dioxide are released
• Weight is increased	• Weight is decreased
• Light is required	• This process can occur in darkness or in light
• Happens within chlorophyll	• This process happens within all living cells
• Oxygen is produced in green organisms	• This process uses oxygen
• ATP is produced with light energy	• ATP is produced when energy is released from sugar

Studying basic plant metabolism helps us understand plant life and other living systems. Inspite of differences among different living beings with respect to respiration and plant photosynthesis, many things are common, especially the nutrient requirements.

Chapter – 10

Carbohydrate Metabolism

Carbohydrate metabolism in the cell is essentially the metabolism of glucose and other substances related to glucose in their metabolic processes. The chemical reactions involved in carbohydrate metabolism represents complex groups, sequences, and cycles of reactions which integrate at various points with the reactions concerned in the metabolism of lipids and proteins. The major types of chemical processes involved in carbohydrate metabolism may be grouped as follows: a) Anabolism of starch – Photosynthesis, b) Catabolism of starch, c) Catabolism of glucose – 1. Glycolysis and TCA cycle and 2. Oxidative pentose phosphate pathway.

Plant cells "burn " glucose for energy. Plants also produce sugars through photosynthesis. This is roughly the reverse of the burning process.

$$\underset{\text{Glucose}}{C_6H_{12}O_6} + \underset{\text{Oxygen}}{6O_2} \underset{\text{Photosynthesis}}{\overset{\text{Burning}}{\rightleftharpoons}} \underset{\text{Carbon dioxide}}{6CO_2} + \underset{\text{Water}}{6H_2O} + 673\text{Cals/ mole}$$

One mole of glucose (180g) on burning releases 673 kcal of heat equivalent to conversion of one litre of water to steam.

Photosynthesis

Green plants use light energy from sun to make carbohydrates. The mechanism by which green plants absorb light and convert carbon dioxide to carbohydrates is what we call photosynthesis. It is the process by which light energy is captured, converted and stored in the simple sugar molecule. This process occurs in chloroplasts and other parts of green organisms. All life on earth depends on the functioning of photosynthesis. Hence it is the most important and primary biochemical process.

This process is composed essentially of two general processes. The first process consists of the absorption of light by the chlorophyll system, which provides energy in the form of activated chemical compounds. The reactions in this case are the so- called light reactions. In the second process carbon dioxide is reduced by the active molecules formed in the light reactions, and the production of carbohydrates takes place. Since these reactions do not require simultaneous illumination, the reactions involved are referred to as the dark reactions. Both the light and dark reactions represent complex interrelated processes under enzymatic control.

Activation of chlorophyll: Chlorophyll on exposure to light absorbs photons and pass into an activated state. The energy of one photon is incorporated into one electron, which flies off into a new orbit more remote from the nucleus. It later falls into the ground state shedding it's extra energy in the form of fluorescent light. The primary function of light in photosynthesis consists in the activation or excitation of chlorophyll molecules. Light-harvesting complexes contain 250 to 400 pigment molecules and are referred to as a photosynthetic unit. There are countless numbers of these units spread throughout the grana of a chloroplast in the green plants.

Activation of chloroplasts: The photochemical activation of chlorophyll is observable in intact chloroplasts. These are the microscopic intracellular organelles in which the chlorophyll is contained. R. Hill in a series of experiments demonstrated the reducing property of chloroplasts. He found that when leaf extracts containing chloroplasts were supplemented with a hydrogen acceptor (2,6-dichloro-phenolindophenol) and illuminated, evolution of oxygen and reduction of the hydrogen acceptor took place. Water undergoes photolysis to release molecular oxygen and nascent hydrogen. The latter is responsible for the reducing property.

$$\underset{\text{Water}}{2H2O} + \underset{\substack{\text{Hydrogen}\\\text{acceptor}}}{2A} \xrightarrow{\text{Light}} \underset{\text{Reduced form}}{2AH2} + \underset{\text{Oxygen}}{O2}$$

Hill reaction

In chloroplasts, the chlorophylls transfer energy in the form of ATP, generated from ADP. This constitutes a mechanism of storing energy , which is needed for a great variety of biological processes. Thus in chloroplasts three things happen 1.creation of reducing atmosphere, 2.photolysis of water and evolution of O_2 and 3. supply of ATP for formation of starch in the later stages.

Light reaction (Hill reaction): Light reaction is the first stage of photosynthesis, the process by which plants capture and store energy from sunlight. In this process, light energy is converted into chemical energy, in the form of the energy-carrying molecules, ATP and NADPH. The light reactions take place on the thylakoid membrane inside a chloroplast. The inside of the thylakoid membrane is called the lumen, and outside of the thylakoid membrane is the stroma, where the light independent reactions take place. The net-reaction of all light reactions in photosynthesis is:

$$\mathbf{2H2O + 2NADP+ + 3ADP + 3Pi \xrightarrow{Light} O2 + 2NADPH + 3ATP}$$

Dark reaction(Calvin cycle): In this pathway, the free energy of cleavage of ~P bonds of ATP, and reducing power of NADPH, are used to fix and reduce CO2 to form carbohydrate. Enzymes and intermediates of the Calvin Cycle are located in the chloroplast stroma, a compartment somewhat analogous to the mitochondrial matrix.

The carbon-fixing reactions are called dark reactions because light does not play a direct role in their functioning. The reactions take place in series outside of the grana in the stroma of the chloroplast. These reactions only occur if the end products of the light reactions are available for use. Depending on the plant involved, the carbon-fixing reactions develop or progress in different ways. The most common type of carbon-fixing reactions in plants is the process called the Calvin cycle. In the Calvin cycle, carbon dioxide from the atmosphere is combined with a 5-carbon sugar—RuBP, (or ribulose bisphosphate). The combined

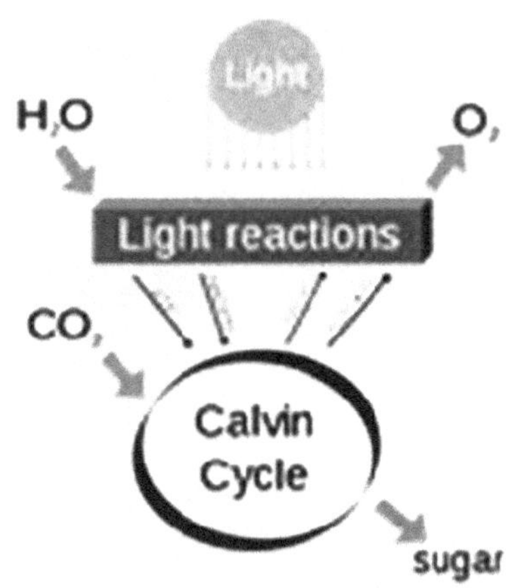

Fig. 1. Photosynthetic process

molecules are converted via several steps into a 6-carbon sugar (glucose). The ATP and NADPH molecules from the light reactions provide the energy and resources for the reactions. Some of the sugars produced are further combined into polysaccharides or are stored as starch within the plant. There are other variations, including the 4-carbon pathway which is usually found in desert plants (C_4 plants).

The Calvin Cycle or the 3-carbon pathway: The Calvin cycle is the most common carbon fixing mechanism in plants. With the assistance of the enzyme rubisco (RuBP carboxylase), six molecules of atmospheric carbon dioxide combine with six molecules of ribulose 1, 5-bisphosphate (RuBP). The result of the first step is six unstable 6-carbon complexes, which immediately split into two 3-carbon molecules of 3-phosphoglyceric acid or 3PGA. This is the first stable compound in photosynthesis. NADPH and ATP from the light-reactions, supply the energy required to convert the 3PGA to 12 molecules of glyceraldehyde 3-phosphates (GA3P), which is a 3-carbon sugar phosphate complex. Finally, of the 12 molecules formed; 10 are restructured into six 5-carbon molecules of RuBP—the starting sugar for the process. The sugars produced can either add to an increase in the carbohydrate content of the plant or they can be used in pathways that lead to the production of lipids and amino acids.

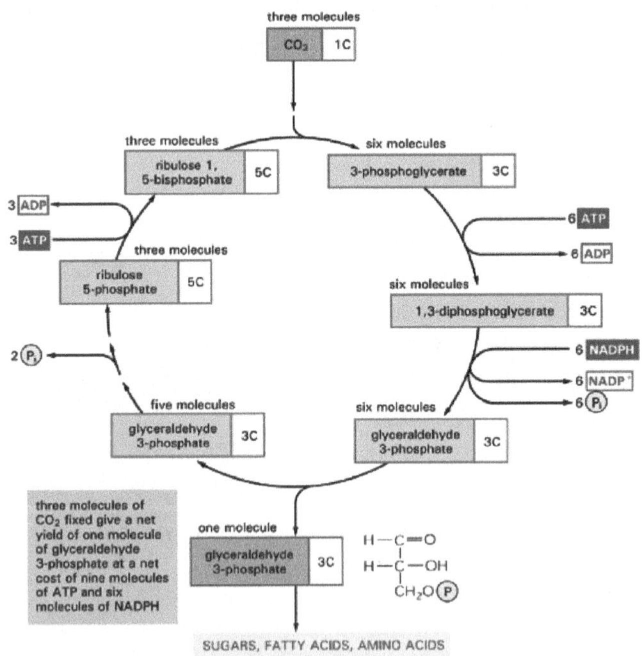

Fig. 2. The Calvin cycle

4-Carbon Pathway—C_4 plants: These plants use a 4-carbon molecule called oxaloacetic acid in place of the 3-carbon 3-phosphoglyceric acid used in the Calvin cycle. Oxaloacetic acid is produced from a 3-carbon compound PEP—phosphoenolpyruvate and carbon dioxide. This process is enzyme mediated and occurs in the mesophyll cells of the leaf. Some species will convert the resulting oxaloacetic acid to aspartic, malic or other acids.

The 4-carbon acids migrate to the bundle sheaths surrounding the vascular bundles, where they are further converted to pyruvic acid and carbon dioxide. In returning to the mesophyll cells and interacting with ATP molecules, the pyruvic acid molecules produce additional PEP. In the bundle sheath cells, the carbon dioxide formed converts into 3PGA and other molecules, by combining with RuBP. The other molecules formed are similar to the ones formed in the Calvin cycle. The C_4 cycle furnishes carbon dioxide to the Calvin cycle in a more roundabout way than the C_3 pathway, but there is an advantage to this extra pathway.

The extra pathway greatly reduces photorespiration in C_4 plants, and this is a good thing because photorespiration is in direct competition with the Calvin cycle and even takes place in the light while the Calvin cycle is functioning. During photorespiration, RuBP reacts with oxygen to create carbon dioxide; in contrast, during photosynthesis RuBP and carbon dioxide are used to form carbohydrates. C_4 plants are able to pick up carbon dioxide in very low concentrations via the mesophyll cells. The Calvin cycle occurs in the bundle sheath where carbon dioxide is readily available. Because of the location, the enzyme rubisco will be in a prime spot to catalyze the reaction between RuBP and carbon dioxide, rather than oxygen. As a result C_4 plants have photosynthetic rates that are two to three times higher than C_3 plants.

There are a few other characteristic features of C_4 plants. They have two types of chloroplasts and an alternate pathway for using carbon dioxide. C_3 plants only have one type of chloroplast and one pathway. PEP carboxylase is found in high concentration in the mesophyll cells which permits ready conversion of carbon dioxide to carbohydrate at lower concentrations than does rubisco (in bundle sheath cells) of the Calvin cycle. The temperature ranges for C_4 plants are much higher than C_3 plants which enable C_4 plants to live well.

CAM Photosynthesis: Crassulacean acid metabolism (CAM) is a modified photosynthetic system that is somewhat similar to C_4 photosynthesis in that 4-carbon compounds are produced during the carbon-fixing reactions. CAM plants accumulate malic acid in their chlorenchyma tissues at night, which is converted back to carbon dioxide during the day. In the daytime, malic acid diffuses out of the vacuoles and is converted to carbon dioxide for use in the Calvin cycle. PEP carboxylase is responsible for converting the carbon dioxide plus PEP to malic acid at night. This modification allows for a greater amount of carbon dioxide to be converted to carbohydrate during the day than would be otherwise under the conditions CAM plants generally grow . CAM plants generally close their stomata during the day in order to reduce water loss. There are more than 20 families that contain CAM plants, including cacti, stonecrops, orchids, bromeliads and many succulents growing in regions of high light intensity.

Catabolism / respiration

Plant respiration is essentially the reverse of photosynthesis. It is a process in which the energy that is stored within organic molecules is released so that metabolic work can take place. Tiny amounts of energy, in the form of ATP is released. It is controlled by enzymes, conducted in all living cells, and releases water and carbon dioxide. Respiration occurs in the mitochondria and cytoplasm of cells. There are mainly three forms of respiration: aerobic—which requires oxygen, anaerobic—which occurs in the absence of oxygen, and fermentation (common in the brewery industry)—which also occurs in the absence of oxygen. In plants, gases passively diffuse into the epidermal cells or through the stomata, into the plant. These gases then meet the moist cellular membranes where water is then moved along diffusion gradients within and between cells. Respiration originates from the glucose molecule. There are several stages of plant respiration.

Glycolysis is the stage in which a 6-carbon glucose molecule is broken down into two, 3-carbon pyruvate molecules. This stage occurs within the cytoplasm. During aerobic respiration pyruvate is used in the Krebs cycle, oxidative phosphorylation, and electron transport chain that occur within the mitochondria. The Krebs cycle, also referred to as the citric acid cycle, happens within the matrix. Oxidative phosphorylation and electron transport chain happen deep within the cristae. During anaerobic respiration, pyruvate is used during fermentation. Lactate formation occurs within bacteria, protist cells, and fungi. Alcohol fermentation occurs within plant cells and yeast.

The thermodynamic efficiency varies between 22 percent and 38 percent in aerobic respiration. It is significantly less in anaerobic respiration. When energy is lost, it is released as heat and some of this heat is used by plants in a few different ways.

Aerobic respiration is the most common form of respiration and cannot be completed without oxygen gas. The controlled release of energy is the main event in aerobic respiration. Certain types of bacteria and other organisms without oxygen gas carry on anaerobic respiration and fermentation. Compared to aerobic respiration the amount of energy released is quite small. The main difference between aerobic respiration and fermentation is in the way hydrogen is released and combined with other substances.

Carbohydrate catabolism deals with the breakdown of different sugars into simpler and smaller units. Carbohydrates are usually taken into cells after they have been converted into monosaccharides. Once inside, the major route of breakdown is glycolysis, where sugars such as glucose and fructose are converted into pyruvate and some ATP is generated. Pyruvate is an intermediate in several metabolic pathways, but the majority is converted to acetyl-CoA and fed into the citric acid cycle (Krebs cycle). More ATP is generated in the citric acid cycle, along with NADH, which is made from NAD^+ as the acetyl-CoA is oxidized. This oxidation releases carbon dioxide as a waste product. In anaerobic conditions, glycolysis produces lactate, through the enzyme lactate dehydrogenase re-oxidizing NADH to NAD+ for re-use in glycolysis. An alternative route for glucose breakdown is the pentose phosphate pathway, which reduces the coenzyme NADPH and produces pentose sugars such as ribose, the sugar component of nucleic acids.

Starch degradation

Starch is hydrolyzed completely to glucose by different enzymes. a) α amylase hydrolyzes α 1-4 glycosidic bonds at random, internally and yields a mixture of dextrins. b) β-amylase hydrolyzes maltosyl residues from non reducing ends and yield only maltose. c) Debranching enzyme(α-1-6 glycosidase) acts at branch points of amylopectin and d) Maltase acts on maltose molecules and yields glucose.

Catabolism of glucose: The metabolism of sugars takes place through their phosphates to pyruvic acid (and lactic acid) in a process called Glycolysis. Pyruvate is the end product of aerobic glycolysis (citric acid or tricarboxylic acid -TCA cycle) whereas lactate is the end product of anaerobic glycolysis. The enzymes of glycolysis are located in the cytoplasm of the cell and are associated into a unit called metabol for efficient functioning. Pyruvic acid is then converted to acetyl CoA, which is oxidized to C02 and H20 in the citric acid cycle.

Glycolysis

The sequence of reactions from glucose to pyruvate is often called as Embden-Meyerhof-Parnas (EMP) pathway. It is also called anaerobic pathway. It is the conversion of glucose to two molecules of lactic acid

$$\underset{\text{(Glucose)}}{C6H12O6} \longrightarrow \underset{\text{(Lactic acid)}}{2\ C3H6O3}$$

During glycolysis glucose is neither oxidized nor reduced. It releases energy. ADP and inorganic phosphate are converted to ATP.

$$2\ ADP + 2\ HPO4' \longrightarrow 2ATP + 2H2O$$

The overall reaction is

$$\underset{\text{Glucose}}{C6H12O6} + 2\ ADP + 2\ HPO4 \longrightarrow \underset{\text{Lactic acid}}{2\ C3H6O3} + 2ATP + 2H2O$$

Glycolysis does not require oxygen gas (O_2) and takes place in the cytoplasm. The glycolytic phase is subdivided into three main steps and several smaller ones. Each step is mediated by an enzyme. A small amount of energy is released and hydrogen atoms are removed from compounds derived from glucose. The main steps of glycolyssis are: a). the glucose molecule goes through several steps and becomes a double phosphorylated fructose molecule. b). The 6-carbon fructose molecule is split into two 3-carbon fragments, each with a phosphate, GA3P. c). Hydorgen, energy and water are removed from the GA3P molecules leaving pyruvic acid.

Glycolysis requires two molecules of ATP to get the process while it produces, four ATP molecules, with a net gain of 2 ATP molecules. The hydrogen ions and electrons that are released are held by an acceptor molecule NAD (nicotinamide adenine dinucleotide). The overall end products of gylcolysis are: 2-ATP molecules, 2-NADH molecules, and pyruvic acid.

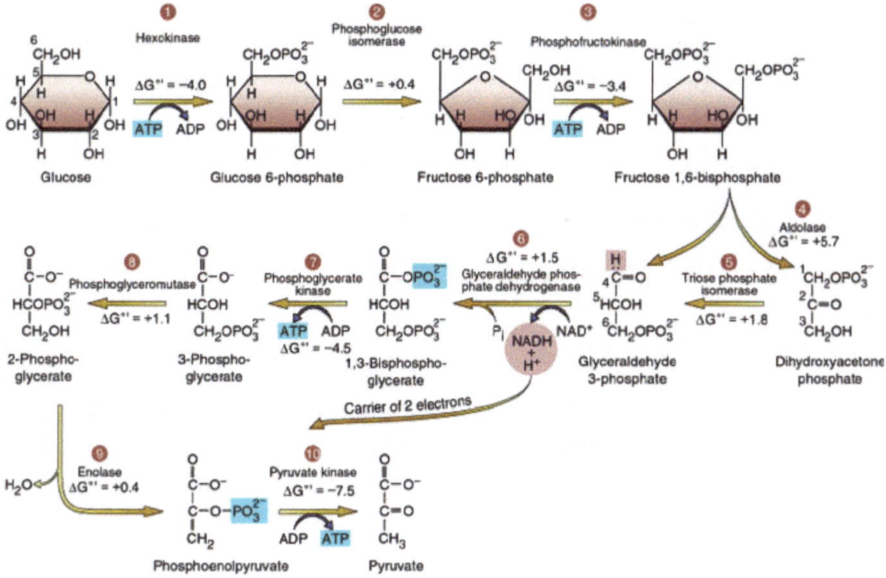

Fig. 3. The glycolysis cycle

Under anaerobic glycolysis, pyruvic acid is converted to lactic acid. In a cell, the molecule which accepts energy from the breakdown of fuel molecules and donates this energy to cell functions is ATP. ATP is synthesized from ADP and inorganic phosphate and 7 kcal/mole must be available to achieve this reaction. This energy is obtained from the catabolism of carbohydrates, fats and proteins.

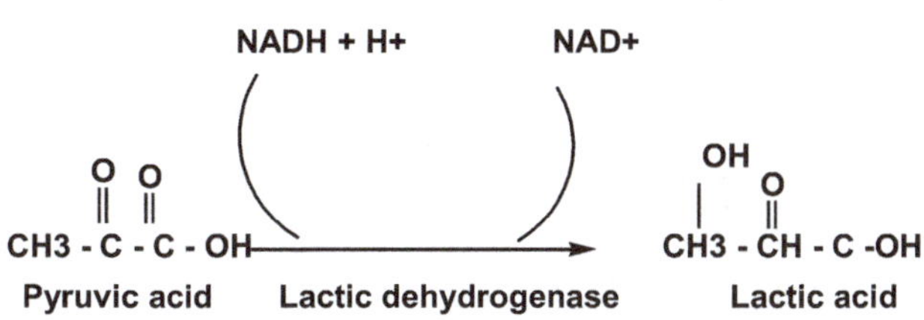

There are two mechanisms by which energy is coupled to the formation of ATP. 1) Substrate level phosphorylation 2) Oxidative phosphorylation. Substrate level phosphorylation occurs when one of the phosphate groups in a metabolic intermediate is transferred to ADP to form ATP. In contrast oxidative phosphorylation couples inorganic phosphate to ADP directly. In the breakdown of glucose, many of the chemical reactions involve the removal of hydrogen atoms from various intermediates. These reactions require a coenzyme, to which the hydrogen atoms are transferred. As a result of such hydrogen transfers, some of the chemical energy in glucose molecule is transferred to the coenzyme molecule. The process of oxidative phosphorylation then uses this energy to synthesize ATP.

The energy release in glycolysis is as follows:

Aerobic glycolysis: Through substrate level phosphorylation = 4 ATPs.

Through oxidative phosphorylation = 6 ATPs. Total = 10 ATPs. Input = 2 ATP

Net energy yield = 8 ATP

The end product of glycolysis, pyruvic acid is converted to acetyl CoA. The reaction is as follows:

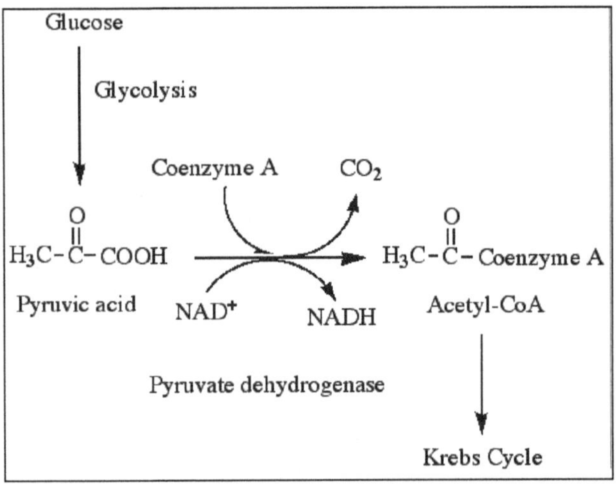

TCA Cycle (Krebs cycle): The enzymes of TCA cycle are located in mitochondria. TCA cycle produces energy efficiently without any consumption. This cycle is fed by pyruvic acid from the anaerobic glycolytic pathway and by lactic acid. Before they enter the Krebs cycle these compounds are converted to their acetyl CoA. This acetyl CoA then transfers its group to oxaloacetic acid forming citric acid. The net reaction of Krebs cycle is the conversion of pyruvic acid to CO_2 and water with the production of 15 molecules of ATP per mole of pyruvic acid.

CH3C=O-COOH + 5/202 +15ADP + 15HPO4 ⟶ 3CO2 +2H2O + 15ATP

NET REACTION OF KREBS CYCLE

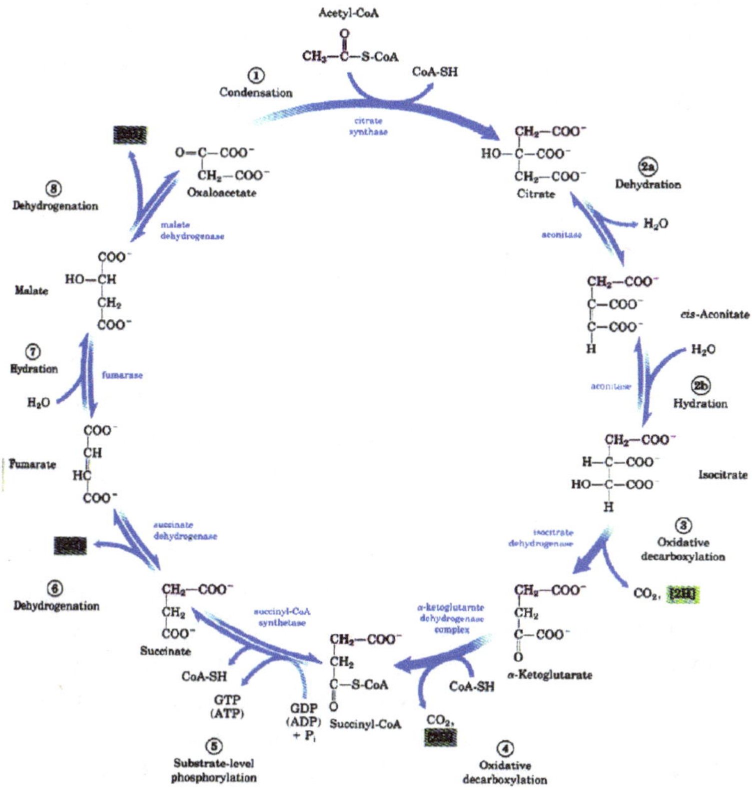

Fig. 4. The citric acid or TCA or Krebs cycle

Most of the pyruvate derived from the PPP (pentose phosphate pathway) and glycolysis is metablised in the TCA cycle. This was first described by Krebs in 1937. The enzymes of the cycle are located in the innermost compartment of the mitochondrion, the matrix. The pyruvate enters the mitochondria from cytosol and is first converted to acetyl CoA in a multistep reaction sequence catalysed by pyruvate dehydrogenase, lipoate acetyl tranferase and dihydro lipoyl dehydrogenase.

Citrate synthase then catalyzes the irreversible formation of citrate from acetyl CoA and the acceptor oxaloacetate. A reversible dehydration followed by hydration catalyzed by aconitase gives rise to cis-aconitate and isicitrate. The formation of alpha ketoglutarate is catalysed by isocitrate dehydrogenase in a decarboxylation reaction dependent on Mg or Mn ions. The next reaction of the TCA cycle is the oxidation of α-ketoglutarate by NAD, accompanied by decarboxylation. The remaining four carbon unit forms the thiol ester, succinyl CoA. The overall reaction is reversible.

The conversion of succinyl COA to succinate is exergonic and is catalyzed by succinyl COA synthase. The conversion of succinate to fumarate is catalyzed by succinate dehydrogenase , an enzyme tightly bound to the mitochondrial membrane. Fumarase then catalyzes the conversion of fumarate to malate in a reversible hydration dehydration reaction. The final step of the cycle is the regeneration of the acceptor oxaloacetate in a reduction reaction catalyzed by malate dehydrogenase.

Pentose phosphate pathway (PPP)

This pathway, also known as the hexose monophosphate pathway produces reduced coenzyme in the form of NADPH, as well as ribose 5-phosphate for incorporation into nucleotides. The net result of the PPP, if not used solely for ribose 5-phosphate production, is the oxidation of glucose 6 phosphate, a 6 carbon sugar, into a 5 carbon sugar. In turn, 3 moles of 5 carbon sugars are converted, via the enzymes of the PPP, back into two moles of 6 carbon sugars and one mole of 3 carbon sugar.

The 6 carbon sugars can be recycled into the pathway in the form of glucose 6 phosphate, generating more NADPH. The 3 carbon sugar generated is glyceraldehyde-3-phosphate which can be shunted to glycolysis and oxidized to pyruvate. This pathway has oxidative and non-oxidative reactions. The oxidative reactions are irreversible, and convert glucose 6-phosphate to five-carbon phospho sugars, with loss of CO2 but concomitant production of NADPH for the reducing power needed for the biosynthesis of fatty acids.

Energetics of pentose phosphate pathway

Glucose 6- phosphate + 12 NADP+ + 7 H2O ⟶ 6CO2 + 12 NADPH+12H+Pi +

The 12 molecules of NADPH on re-oxidation to NADP+ by NAD+ give rise to 12 units of NADH which through oxidative phosphorylation yields about the same number of ATP molecules (12 NADH = 36 ATP molecules).

Significance of pentose phosphate pathway

Apart from the energetics point of view, this pathway has other significant metabolic functions. 1. The NADPH which is generated through this pathway is required in reductive biosynthesis of fatty acids and steroids. 2. It supplies ribose needed for nucleic acid biosynthesis. 3. The ribose-5-phosphate which is produced as an intermediate is utilized for photosynthetic formation of glucose from CO_2 in some plants. 4. It also supplies erythrose-4-phosphate required for the biosynthesis of phenylalanine, tyrosine and tryptophan. 5. In humans, the shunt pathway is operative in liver where there is considerable biosynthetic activity and in red blood cells where NADPH is required to maintain high levels of reduced glutathione, a biological reducing agent. 6. This pathway also makes possible for inter-conversion of various three, four, five, six and seven carbon sugars and connects all such sugars metabolically with the glycolytic sequence.

The PPP differs from glycolysis in a number of ways. Here the oxidation takes place in the first reactions with the production of reduced NADP. Carbon dioxide is evolved and no energy is trapped. Enzymes of the glycolysis and the PPP pathways are soluble in the cytosol. Both pathways operate in the plants and the PPP pathway contributes to about 30% of the total carbohydrate oxidized.

Electron transport chain (ETC) in mitochondria

Electron transport chain is responsible for release of energy and is dumped into the system. When the electrons reach the end of the chain they pick up an oxygen and water is released. ATP is produced by oxidative phosphorylation during the action of the electron transport chain. This occurs essentially like chemiosmosis. Oxidative phosphorylation is central to metabolism because the energy of ATP is used in the synthesis of nucleic acids, proteins and complex lipids.

Reduced cofactors like NADH and FADH can be oxidized back to NAD and FAD and in the process energy liberated is used to form ATP molecules at the electron transport chain. This chain is located in the mitochondrion on the inner membrane. All the enzymatic steps in the oxidative degradation of carbohydrates, fats and amino acids in aerobic cells converge at the final stage of cellular respiration in which electrons flow from the catabolic intermediates to molecular oxygen. This yields energy for the generation of ATP from ADP + Pi. ATP synthesis in mitochondrion and chloroplast is based on a hypothesis introduced by Peter Mitchell in 1961. It is now widely accepted and is called chemi-osmotic theory.

Mitochondrial electron flow: Here electrons pass from lower to higher reduction potential. Many of the proteins involved in electron flow are embedded in the inner mitochondrial membrane. They are organized into 4 respiratory complexes – I, II, III, and IV. Electrons are carried from I and II to III by Coenzyme Q (Ubiquinone) and from III to IV by

Cytochrome c. Most of the electrons entering into mitochondrial respiratory chain arise from the action of dehydrogenases which collect electrons from oxidative reactions of Glycolysis, TCA cycle, β- oxidation and amino acid catabolism.

1. Complex I contains FMN and iron sulphur clusters also called NADH dehydrogenases. Complex I catalyses the oxidation of NADH by Coenzyme Q (Co Q)

NADH + Co Q (oxidized) ⟶ NAD + + CoQ (reduced)

2. Complex II contains Succinate dehydrogenase and 3 small hydrophobic units called SDH complex. Complex II catalyses the oxidation of FADH2 by Co Q. It passes the electrons from succinate to Co Q.

FADH2 + Co Q (oxidized) ⟶ FAD + + Co Q (reduced)

3. Complex III contains 2 cytochrome b molecules and 1 cytochrome c called ubiquinone – cytochrome c oxidoreductase. Complex III catalyses the oxidation of CoQ by Cytochrome c

Co Q (reduced) + Cyt c ⟶ Co Q (oxidized) + Cyt c (reduced)

4. Complex IV contains 2 cytochrome a molecules and 2 copper atoms called as cytochrome oxidase. Complex IV catalyses the oxidation of Cytochrome c by O_2 (the terminal electron acceptor of ETC)

Cyt c (reduced) + ½ O2 ⟶ Cyt c (oxidized) +H2O

The proton consuming steps are located towards the matrix side and the proton releasing steps are located towards the inter membrane space. Protons accumulate in the inter membrane space whenever electrons flow through the chain. This proton gradient is dispersed by back flow of protons through ATP synthase. As a result, 3 ATP molecules are generated when NADH is oxidized and 2 ATP molecules are generated when FADH is oxidized. The terminal electron acceptor in ETC (electron transport chain) is O_2 or the common electron sink is O_2. Cytochromes are iron containing electron transfer proteins of inner mitochondrial membrane.

Electron transport chain in chloroplasts: The cells of all plants and photosynthetic algae contain chloroplasts, which produce ATP and

NADPH using the energy of sunlight. The electron transport chain in chloroplasts is contained in two extremely complex trans-membrane structures, *Photosystem II* (PS II) and *Photosystem I* (PS I). PS II and PS I are linked by a trans-membrane proton pump, cytochrome *b6f*, which is similar to mitochondrial *Complex III*. The overall process is the transfer of electrons from water to NADPH via a transmembrane proton pump:

H2O → PS II → plastoquinone → *b6f* → plastocyanin → PS I → NADPH

The resulting trans-membrane proton gradient is used to make ATP via ATP synthase. NADPH is used by the Calvin cycle to make inorganic molecules from CO2 and nitrates

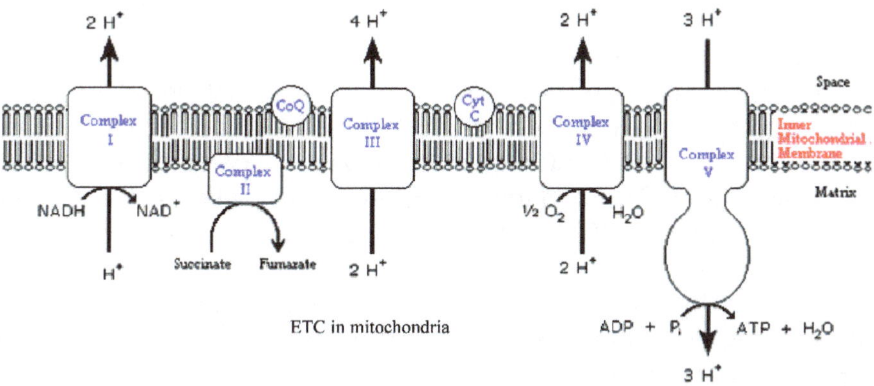

Fig. 5. Electron transport chain in mitochondria

Photosystems

The two types of photosynthetic units in most chloroplasts are what constitute photosystem I and photosystem II.

Photosystem I: contains photosynthetic units with 200 or more molecules of chlorophyll a, small amounts of chlorophyll b, protein saddled carotenoid pigment, and a pair of specialized reaction-center molecules of chlorophyll called P_{700} (absorption peak, ie.,700nm). All pigments in a photosystem are capable of absorbing photons. Iron-sulphur complexed proteins initially receive electrons from P_{700} and serve as primary electron acceptors for the unit.

Photosystem II: contains chlorophyll a, protein saddled beta-carotene, a small amount of chlorophyll b and special pair of reaction-center molecules of chlorophyll a otherwise called, P_{680} (absorption peak, ie.,680nm). The photosystem has a primary electron acceptor called pheophytin or Pheo. PS II is an extremely complex, highly organized transmembrane structure that contains a *water-splitting complex*, chlorophylls *a* and *b*, a *reaction center* (P680), pheophytin (a pigment similar to chlorophyll), and two quinones. It uses the energy of sunlight to transfer electrons from water to a mobile electron carrier in the membrane called *plastoquinone*:

$$H_2O \rightarrow P680 \rightarrow P680^* \rightarrow \text{plastoquinone}$$

Plastoquinone, in turn, transfers electrons to *b6f*, which feeds them into PSI.

Photolysis

A photon of light strikes the photosystem II reaction-center, the P_{680} molecule to be exact near the inner surface of a thylakoid membrane. The received light energy excites an electron which is an unstable reaction and thus most of the energy is lost to heat. Up to four photons at a time can strike the P680 molecule, however, it can only accept one electron at a time. The molecule of pheophytin picks up the excited electron, which then crosses the thylakoid membrane and is passed along to another acceptor called plastoquinone or Pq near the outside surface of the thylakoid membrane. Protein Z extracts electrons from water and replaces the ones lost by the P680 molecule. Protein Z contains manganese which is required in order to split water molecules. Simultaneously, as two water molecules are split, a molecule of oxygen and four protons are produced. This enzyme-mediated water splitting process is called photolysis.

Photophosphorylation

Pq, the acceptor molecule, releases the excited electron into an electron transport system. The transport system moves electrons extracted from water temporarily to a high-energy storage molecule called nicotinamide adenine dinucleotide phosphate ($NADP^+$). $NADP^+$ is an electron acceptor for photosystem. The transport chain is essentially iron-containing pigments, cytochromes, a copper containing protein called plastocyanin and other electron transferring molecules. As electrons are passed through the chain, ATP molecules are assembled from ADP and phosphate in a process called photophosphorylation.

A similar series of events occur in photosystem I. After a photon of light strikes a P_{700} molecule, the resulting excited electron is passed along to an iron-sulphur molecule Fe-S which in turn passes it to another acceptor molecule ferrodoxin, (Fd). The ferrodoxin molecule releases the electron to a carrier molecule called flavin adenine dinucleotide (FAD) and then eventually on to $NADP^+$. A reduction occurs and NADP+ becomes NADPH. Electrons from photosystem II and the activities of the electron transport system replace any electrons removed from the P_{700}molecule. Because the electrons move in one direction, the movement of electrons from water to photosystem II to photosystem I to $NADP^+$ are considered noncyclic electron flow.

PS I accepts electrons from plastocyanin and transfers them either to NADPH (*noncyclic electron transport*) or back to cytochrome *b6f* (*cyclic electron transport*):

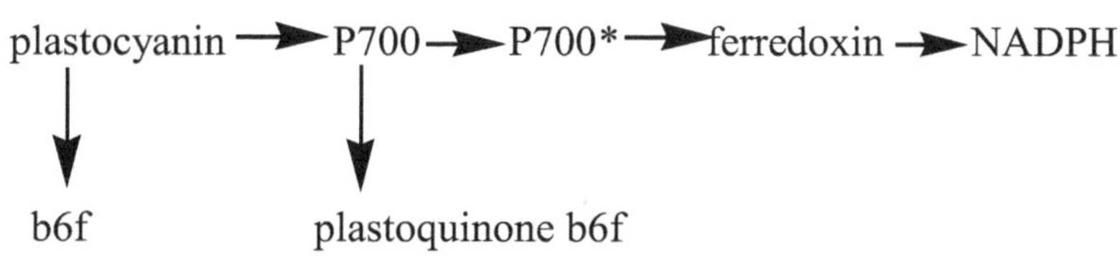

There are two different pathways of electron transport in PS I. In *noncyclic electron transport*, ferredoxin carries the electron to the enzyme ferredoxin NADP + oxidoreductase that reduces NADP + to NADPH. Alternately, in *cyclic electron transport*, electrons from ferredoxin are transferred (via plastoquinone) to a proton pump, cytochrome *b6f*. They are then returned (via plastocyanin) to P700. NADPH and ATP are used to synthesize organic molecules from CO_2. The ratio of NADPH to ATP production can be adjusted by adjusting the balance between cyclic and noncyclic electron transport.

Factors affecting respiration

Temperature: The higher the temperature the faster respiration. This occurs till the activity of the enzyme is not affected. Energy from sugar is released faster as the rate of respiration increases which results in a net weight loss. Plants offset the weight loss by increasing photosynthetic production of sugar.

Water: Enzymes generally operate in the presence of water, and reduced water in a plant will reduce the rate of respiration. Seeds usually have a water content of less than 10%, while mature living cells usually are in excess of 90% water. Seeds keep better if they are kept dry as the respiration rate remains quite low. However, if a seed comes into contact with water and via imbibition swells, the respiration rate increases. The temperature could increase to the point of killing the seeds. Spontaneous combustion can occur from the respiration generated heat when a fungi or bacterium is permitted to grow on wet seeds.

Oxygen: Oxygen is an important regulator of respiration. If oxygen is drastically reduced, respiration may drop off to the point of retarding the growth and may lead to death. Low oxygen concentrations can lead to the onset of fermentation processes.

Chapter – 11

Lipid Metabolism

The importance of lipids have already been elaborated. They play a major role as source of energy and as components of membranes which have implications in plant growth regulation. The complete metabolism of fat leads to oxidation, the final products being CO_2 and water. The burning of one mole of palmitic acid gives 2340 kcal. The molecular weight of palmitic acid is 256. This works out to approximately 9 kcal/g of palmitic acid. This energy is twice that derived from either sugars or proteins. Fats and oils are thus better and more efficient forms of storage of energy in living beings.

Anabolism of saturated fatty acids

Fatty acid synthesis is characteristic of all living organisms. Multi-enzyme complexes referred to as type I fatty acid synthases are essential for fatty acid synthesis. Acetyl coenzyme A is a metabolite of importance to various pathways in living beings. A pool of acetyl coenzyme A exists which is regularly replenished and depleted. An enzyme acetyl coenzyme A synthase catalyzes the formation of thiol ester from acetate.

$$\underset{\text{ACETATE}}{\text{CH3COO-}} + \text{HSCOA} \longrightarrow \underset{\text{ACETYL COA}}{\text{CH3COS-COA}} + \text{OH}$$

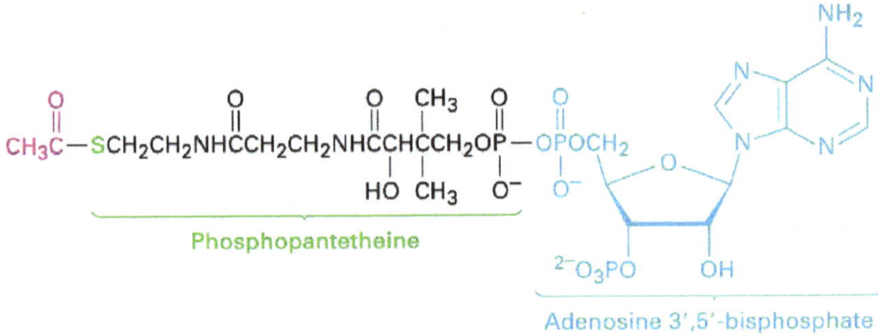

Acetyl CoA—a thioester

The acetyl group in acetyl CoA is linked to the sulfur atom of phosphopantetheine, which is itself linked to adenosine 3',5'-bisphosphate.

The first reaction of the fatty acid synthesis sequence in plants is the carboxylation of acetyl CoA to malonyl coenzyme A in the presence of the enzyme acetyl CoA carboxylase.

ACETYL CoA + CO2 +ATP ⟶ MALONYL CoA + ADP + Pi

Malonyl Co A is also formed from oxaloacetate,

OOC-CO-CH2COO $\xrightarrow{1/2\ O2}$ OOC-CH2-COO- +CO2

OXALO ACETATE MALONATE

One molecule of acetyl CoA and one molecule of malonyl CoA are converted to their corresponding ACP derivatives in the presence of the enzymes trans-acylases. Malonyl CoA condenses with an acetyl CoA to yield aceto acetyl CoA and CO_2 is released. This process is carried out with the help of ACP (acyl carrier protein)- a multi enzyme complex. The two carbon acetyl residues are sequentially condensed with aceto acetyl CoA on the ACP complex until a 16C fatty acid (palmitic acid) is formed. One further acetyl moiety is added to this by an elongation process to yield an 18:0 ACP complex, from which 18:0 (stearic acid) is released.

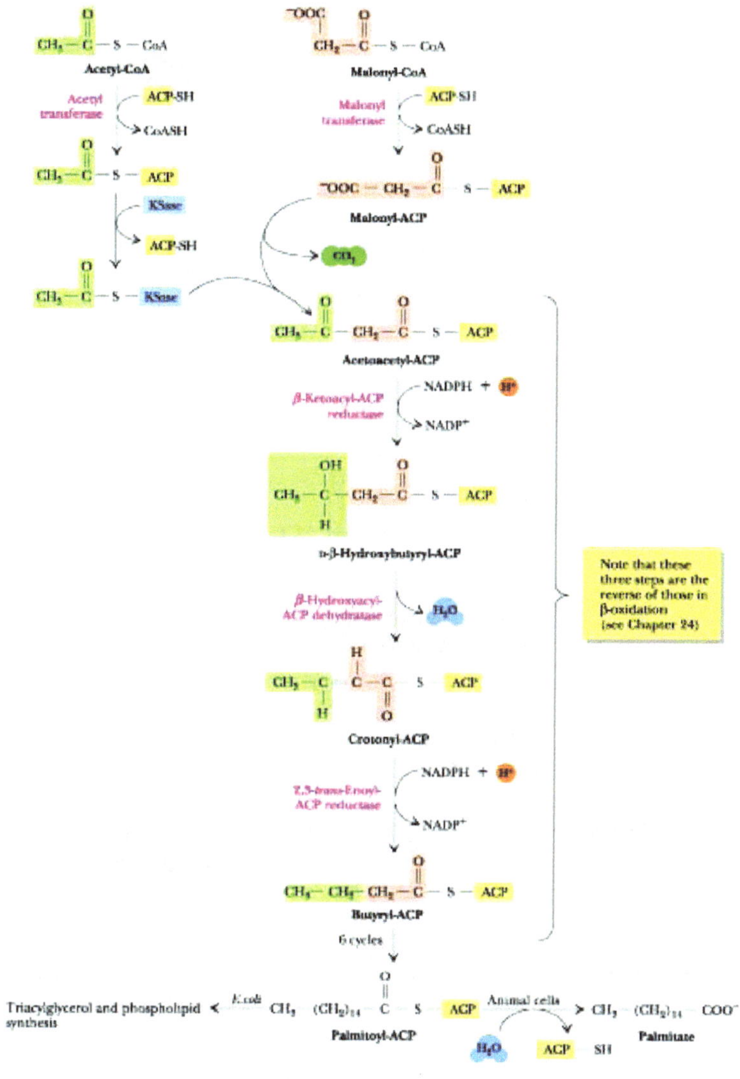

Fig. 1. Pathway of fatty acid biosynthesis

Anabolism of unsaturated fatty acids

The fatty acid synthesized generally is palmitate which is a 16:0 fatty acid. Unsaturation of fatty acids occurs in both the mitochondria and endoplasmic reticulum in the presence of enzymes called f*atty acyl-CoA desaturases*. Aerobic desaturation of stearic acid by stearyl desaturase produces oleic acid (18:1). Further desaturation yields linoleic acid (18:2).

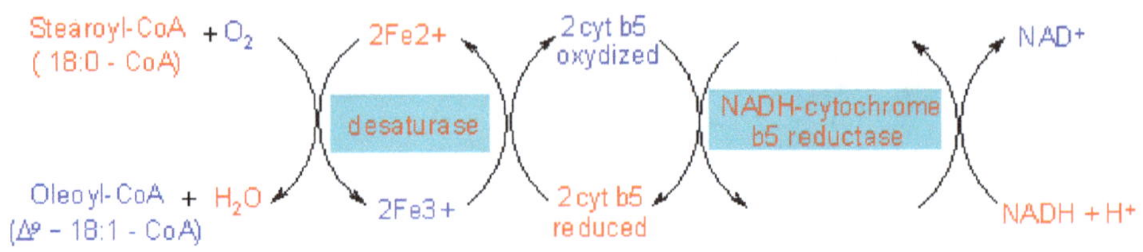

Fig. 2. Desaturation of stearic acid

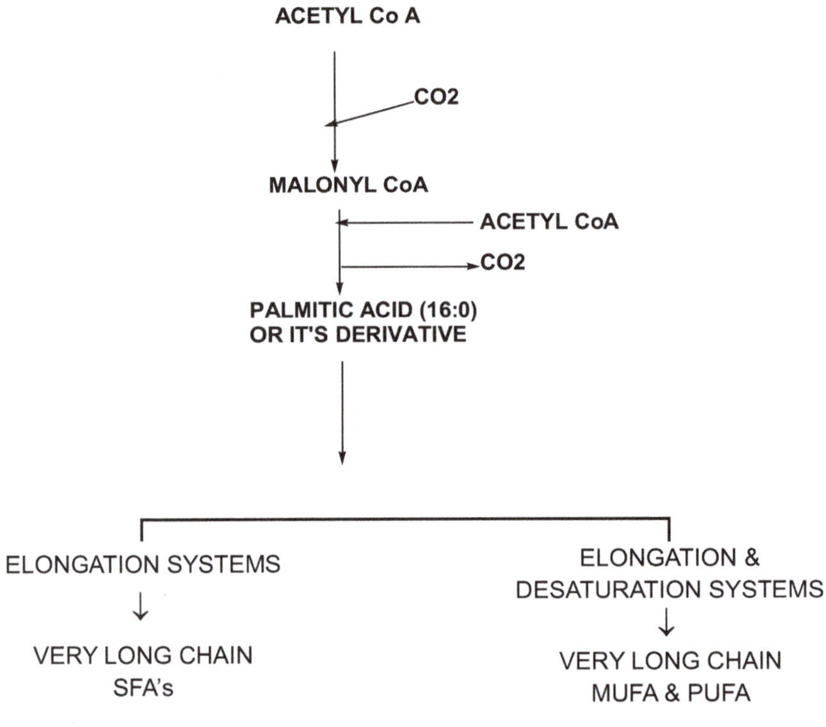

Fig. 3. Biosynthesis of saturated and unsaturated fatty acids

Anabolism of triacyl glycerol

The most important route to triacylglycerol biosynthesis is the Kennedy pathway. Glycerol is asymmetrically phosphorylated to form L-alpha glycerophosphate. This resaction is catalyzed by glycerokinase which has been demonstrated in the germinating seeds. The next step involves condensation of two acyl CoA molecules with L- alpha glycerophosphate to form L- alpha phosphatidic acid. The acylation is followed by dephosphorylation of the phosphatidic acid to form a diglyceride. The condensation of a further molecule of an acyl CoA with the free hydroxyl group of the diglyceride to form a triglyceride.

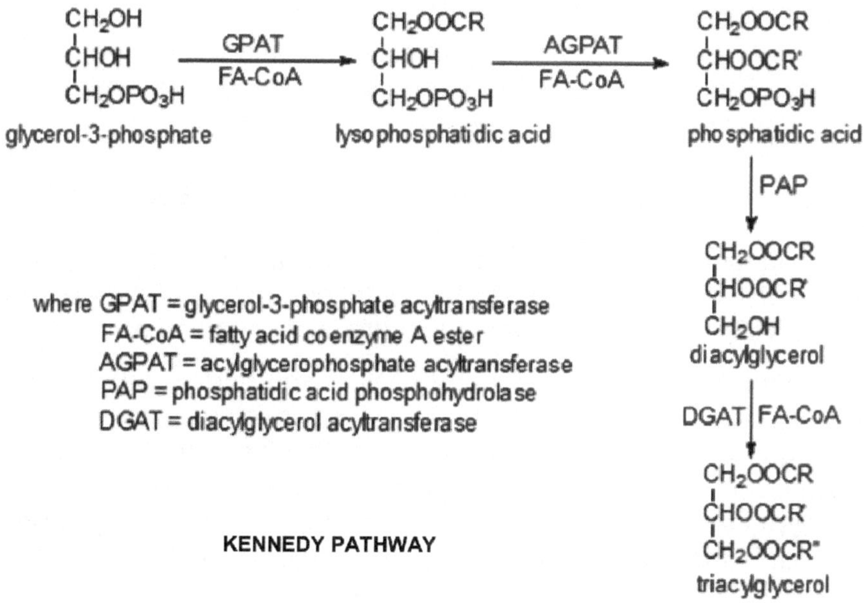

Fig. 4. Kennedy pathway of triacylglycerol biosynthesis

Catabolism of lipids

Fats are metabolized in a multistep process that produces energy. The first step in lipid catabolism is the hydrolysis of the lipid in the cytoplasm to produce glycerol and fatty acids. The glycerol is phosphorylated to glycerol phosphate and then is oxidized to dihydroxy acetone phosphate.

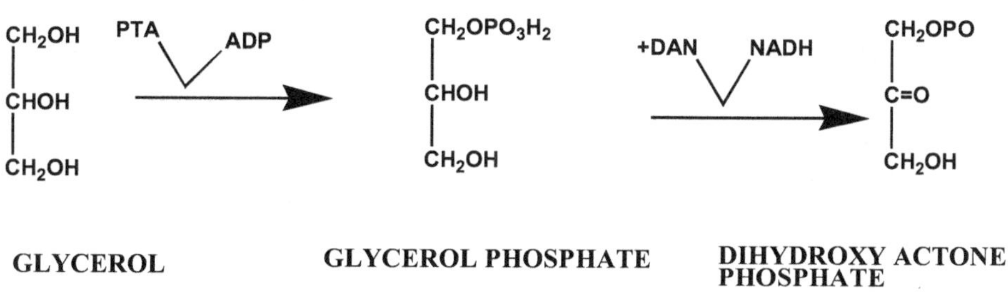

Catabolism of fatty acids

In the mitochondria, fatty acids are broken down by various types of oxidations such as α, β, and γ oxidation. A step by step degradation of fatty acids take place in the three oxidation processes.

Alpha oxidation: Alpha oxidation occurs in plants where odd number of carbon containing fatty acids are present. A fatty acid containing n carbon atoms gives rise to n-1 carbon containing fatty acid.

R-CH2-CH2-CH2-COOH ⟶ R-CH2-CH2-COOH

Alpha oxidation produces 2-hydroxy fatty acids which are important constituents of sphingolipids. Alpha oxidation is as important as beta oxidation which is the most prevalent fatty acid oxidation process.

β-oxidation: In β-oxidation, the fatty acid (even numbered) is broken down to release acetyl-CoA. The process involves 4 main steps: dehydrogenation, hydration, oxidation and thiolysis. In the first step, oxidation (dehydrogenation) takes place, acyl CoA is converted to trans Δ^2–enoyl CoA. In the second hydration step, 3-hydroxyacyl CoA is produced. In the third oxidation step, 3-ketoacyl CoA is formed with the help of NAD^+. In the next step of thiolysis by a second CoA molecule, an acyl CoA is reduced by two carbon atoms. The process repeats until the fatty acid has been completely degraded to acetyl-CoA. Each round of β-oxidation yields 1 molecule of acetyl CoA and requires 1 molecule of NAD+ (nicotinamide adenine dinucleotide) and 1 molecule of FAD+ (flavin adenine dinucleotide). Hence each round of β-oxidation releases 5 ATP molecules. For example, the β-oxidation of a C16 fatty acid generates 8 molecules of acetyl CoA and 7 molecules of NAD+ and FAD+. The intermediates of beta oxidation do not accumulate and thus short chain fatty acids are formed by biosynthesis rather than the catabolic process.

Energy yield: The total degradation of palmitoyl CoA (16:0) requires seven rounds of degradation producing 7 x 5 = 35 ATP molecules. Eight molecules of acetyl CoA are produced and hence another 8 x 12 = 96 ATP. The total ATP yield is 35 + 96 = 131. However, 2ATPs are lost for activation of palmitate. Hence the actual yield of ATPs is 129. The yield of ATPs is slightly reduced for unsaturated fatty acids.

In plants the acetyl COA resulting from β -oxidation is concerned in two processes. In the mitochondria, it is oxidized to CO_2 by the TCA cycle. This process releases energy in the form of ATP. In the organelles known as glyoxysomes, acetyl COA is converted to glucose through the glyoxylate pathway. The succinate generated by this pathway is transported to the mitochondria where it joins the TCA cycle.

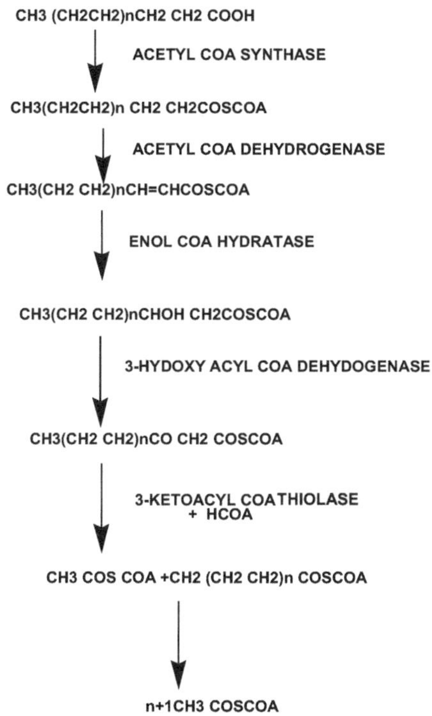

Fig 5. β-Oxidation of fatty acids

Omega oxidation: It occurs in the endoplasmic reticulum rather than the mitochondria, the site of beta-oxidation. The omega carbon in a fatty acid is the carbon farthest in the alkyl chain from the carboxylic acid. In the ω-oxidation pathway, this carbon is progressively oxidized first to an alcohol and then to a carboxylic acid, creating a molecule with a carboxylic acid on both ends.

$$\text{R-CH2-CH2-CH2-COOH} \longrightarrow \text{COOH-CH2-CH2-CH2-COOH}$$

Glyoxylate cycle: In plants, the glyoxylate cycle occurs in special peroxisomes which are called glyoxysomes. This cycle allows seeds to use lipids as a source of energy to form the shoot during germination. The lipid stores of germinating seeds are used for the formation of the carbohydrates that fuel the growth and development of the organism. The two enzymes which regulate this cycle are isocitrate lyase and malate synthase.

Chapter – 12

Amino Acid Metabolism

Plants can synthesize all the 20 amino acids, which include essential and non-essential amino acids. They can be degraded by removal of the amino group and conversion of the remaining skeleton to useful metabolic intermediates which participate in the synthesis of glucose and other ketone bodies.

Biosynthesis of amino acids

The biosyntheses of amino acids have an important feature. Their carbon skeletons come from intermediates from glycolysis, pentose phosphate pathway and citric acid cycle etc. The biosynthetic pathways depend on the intermediates from which they are derived. The primary amino group normally comes from the transamination of glutamate. Glutamine synthetase is a central control point in nitrogen metabolism, since glutamine is the amino group donor in the formation of many biosynthetic products as well as being a storage form of ammonia. Glutamate is the precursor of proline, ornithine, and arginine. Serine, cysteine, and glycine are derived from 3-phosphoglycerate. Glycine is synthesized from serine by removing CH2OH group. Tyrosine is synthesized from phenylalanine. Asparagine and glutamine are amides that are synthesized from aspartate and glutamate.

Tetrahydrofolate (THF) functions to transfer C1 units in several oxidation states. Most reactions require NADPH/NADH. THF is composed of three units: 2-amino-4-oxo-6-methylpterin, p- aminobenzoic acid and glutamates. THF is derived from folic acid by a two-stage reduction. The C1 units carried by THF are converted to the amino acid namely, methionine. Plants and microorganisms can synthesize all the 21 aminoaids, essential and non- essential. Humans cannot synthesize the essential amino aids. Hence they depend on plants for their essential amino acid needs.

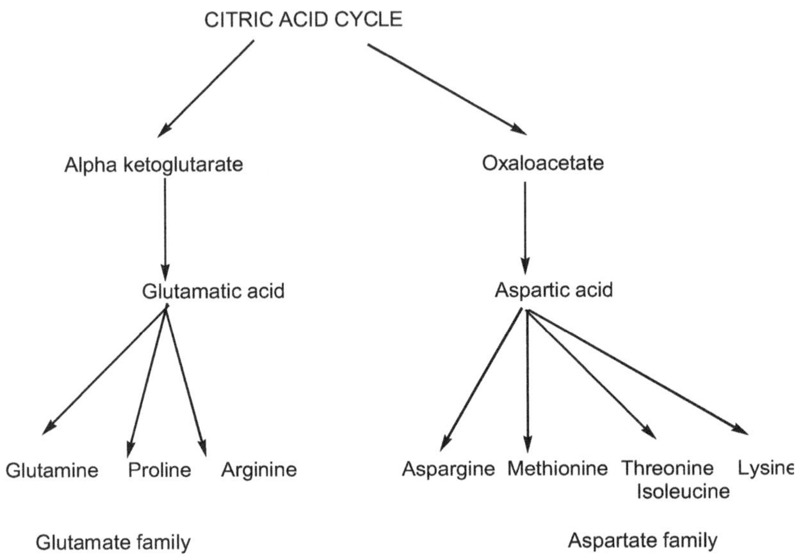

Fig. 1. Biosynthesis of glutamate and aspartate family amino acids

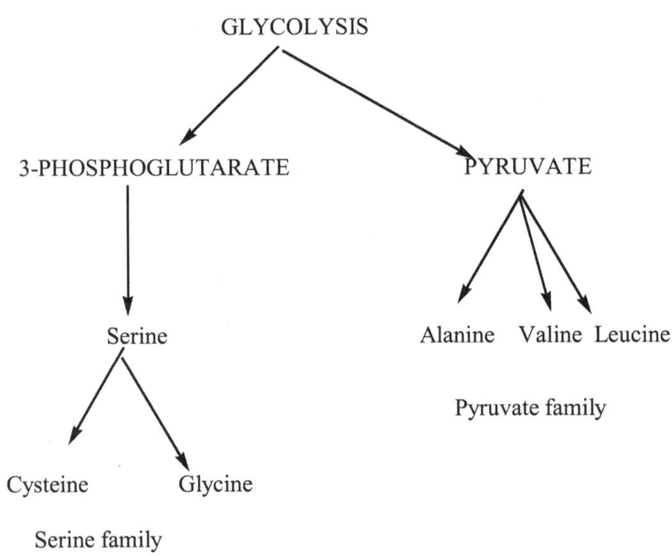

Fig. 2. Biosynthesis of serine and pyruvate family amino acids

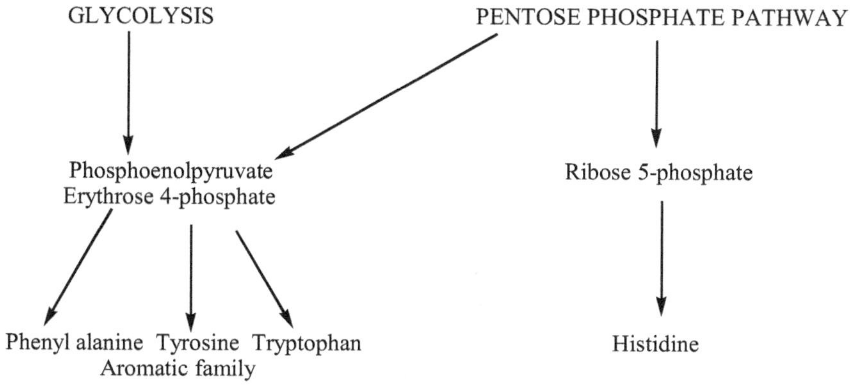

Fig. 3. Biosynthesis of aromatic amino acids

Amino acid degradation

Generally amino acids do not get stored. Hence they get continually degraded. The amino acids undergo catabolism in the following ways. a. Transamination b. Oxidative deamination. c. Non- oxidative deamination and d. Decarboxylation

Transamination: This reaction involves the transfer of an amino group from a donor amino acid to an acceptor α-keto acid to yield the α-ketoacid of the donor amino acid and the amino acid of the original α-keto acid. The reaction is catalyzed by an enzyme called aminotransferase or transaminase that requires PLP as co-enzyme for its activity.

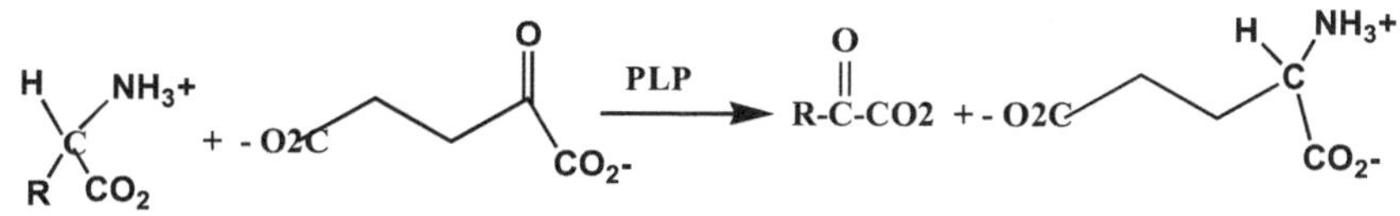

Amino acid **Alpha keto glut aric** **Alpha keto acid** **Glutamate**

Oxidative Deamination: Amino group from amino acid is oxidatively removed in this process. Glutamate dehrogenase enzyme catalyses the reaction.

$$\text{Glutamate} \xrightarrow[\text{dehydrogenase}]{\text{Glutamate}} \text{Ketoglutarate}$$

Non oxidative deamination: Non oxidative deamination takes place in the absence of water and the corresponding unsaturated dicarboxylic acid is formed

$$\text{Aspartic acid} + H2 \longrightarrow \text{Fumaric acid} + NH3$$

Decarboxylation: In decarboxylation reactions, the CO2 from the corresponding amino acids are removed and biologically active amines are formed.

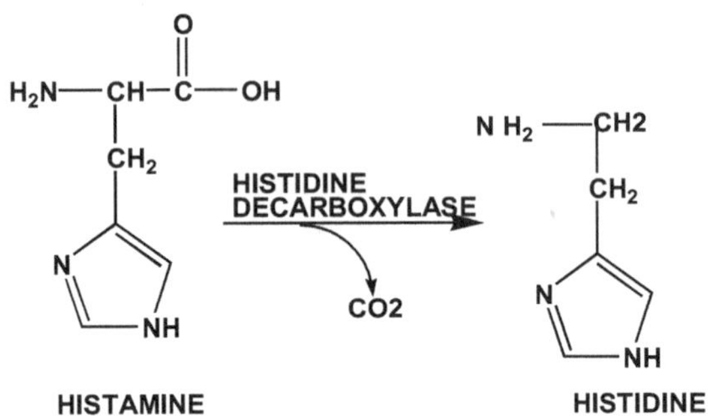

Amino acid carbon skeletons: The remainder of the amino acid is referred to as the "carbon skeleton". Depending on the particular amino acid being catabolised, its carbon skeleton will be converted to acetyl CoA, or pyruvate and or a citric acid cycle intermediate. Those carbon skeletons which end up as acetyl CoA are committed to energy production. They get oxidised via the citric acid cycle.

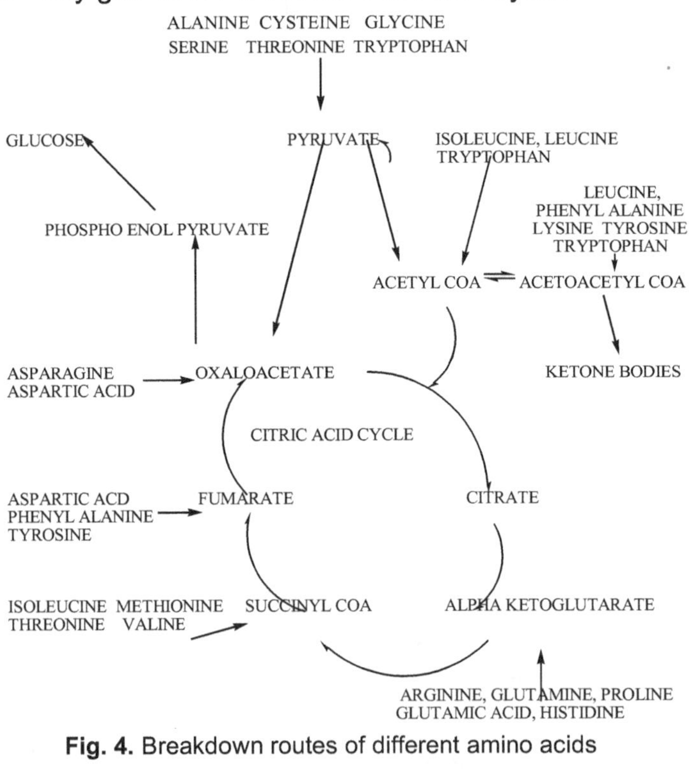

Fig. 4. Breakdown routes of different amino acids

Ammonia assimilation

The ammonia released during metabolism of amino acids is toxic to the cells if it is not incorporated into synthetic processes. There are three reactions that can catalyze the incorporation of the nitrogen atom as NH_3 into organic compounds. These reactions are those catalysed by glutamic dehydrogenase, glutamine synthetase and carbamyl phosphate synthetase. In the animal systems ammonia gets converted to urea.

$$CO2 + 2ATP \xrightarrow[\text{synthetase}]{\text{Carbomyl phosphate}} \text{Carbamyl phosphate } 2ADP + H_3PO^{=}_4$$

$$NH_3 + \text{glutamate} + ATP \xrightarrow[Mg^{2+}]{GS} \text{glutamine} + ADP + P_i \quad (1)$$

$$\text{glutamine} + \alpha\text{-ketoglutarate} + NADPH \xrightarrow{GOGAT} 2\ \text{glutamate} + NADP^+ \quad (2)$$

$$NH_3 + \alpha\text{-ketoglutarate} + NADPH \xrightarrow{GDH} \text{glutamate} + NADP^+ \quad (3)$$

Chapter – 13

Protein Metabolism

Amino acids serve as the building blocks of proteins and as precursors of many other important biomolecules such as hormones, purines, pyrimidines, porphyrins and some vitamins. They also serve as sources of energy when present in excess amounts and when they are used as fuel, amino acids undergo deamination reactions. The remaining carbon skeleton will either convert itself into glucose or undergoes oxidation to CO_2 via TCA cycle.

Central dogma: The Central dogma of molecular biology, which has been formulated as a simple linear progression of information from DNA to RNA to Protein, is summarized in the illustration. The replication process on the left consists of passing information from a parent DNA molecule to daughter molecules. The middle transcription process copies this information to a mRNA molecule. Finally, this information is used by the chemical machinery of the ribosome to make polypeptides.

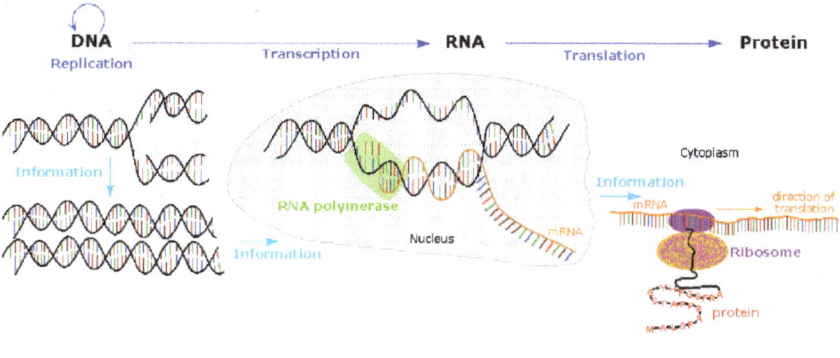

Fig. 1. Progression of information from DNA through RNA to protein

The central dogma consists of 3 major processes in the preservation and transmission of genetic information. a) Replication: It is the process by which each strand of parent DNA molecule is copied to form two identical daughter DNA molecules. b) Transcription: It is the process by which genetic message in DNA is transcribed (copied) into the form of mRNA. c) Translation: It is the process by which the information transcribed from DNA to mRNA directs the specific amino acid sequence enabling specific protein synthesis. The flow of genetic information as proposed by Crick is the central dogma of molecular biology.

The machinery of protein biosynthesis consists of DNA, mRNA, tRNA, ribosomes, and a large number of enzymes and cofactors. The whole machinery operates in the cytoplasm and is closely connected to the endoplasmic reticulum.

Ribosomes: They act as a platform on which the protein synthetic machinery is assembled. They have two sites- amino acyl or A site and peptidyl or P site. The initiating codon AUG is positioned at P site. During elongation, all the incoming amino acyl tRNAs bind to A site. Ribosomes of prokaryotes have sedimentation coefficient of 70S. During translation, there is continuous dissociation of 70S ribosome into 50S and 30S subunits. mRNAs and tRNAs cannot bind directly to 70S subunit but first bind to 30S subunit which then combines with 50S subunits. Eukaryotes have 80S sedimentation coefficient (40S and 60S). S is Svedberg unit. It is a measure of sedimentation cofficient. The S units of protein subunits are not additive.

mRNA: The key component of translation is mRNA. It carries the genetic information from DNA to the site of protein biosynthesis, namely, the ribosome. mRNA contains codons. Codon is the unit that codes for a given amino acid and consists of 3 nucleotides.

Genetic code: DNA replication is simply a complementary base pairing exercise. But the translation of the four letter (bases) alphabet code of RNA to the twenty letter (amino acids) alphabet of protein literature is the important route. Clearly, there could not be a direct one-to-one correlation of bases to amino acids, so the nucleotide letters must form short words or codons that define specific amino acids.

The features of genetic code are *a)* The code/codon is a triplet (as singlets and doublets are insufficient to code 20 amino acids), *b)* The genetic code is universal. The same code dictionary is used by all organisms (both prokaryotes and eukaryotes), *c)* The genetic code is comma less. It is a series of codons without space or punctuations. *d)* It is non-overlapping. The nucleotide sequence of 2 consecutive amino acids do not overlap, *e)* It is degenerate. With the exception of codes for methionine and tryptophan, all the amino acids are encoded by more than one codon.

tRNA: tRNA has a clover leaf structure. The molecule is asymmetrically folded to yield a compact structure with anticodon arm at one end and amino acid arm at the other end. Anticodon is the nucleotide triplet in tRNA. It is present in the central petal of tRNA. Anticodon is complimentary to the corresponding codon in mRNA. tRNAs are small in size consisting of 70-90 nucleotides.

Translation (Protein biosynthesis)

Protein synthesis is the most complex biosynthetic mechanism. Despite the complexity, proteins are made at a high speed. A polypeptide chain of 100 amino acid residues is synthesized in a few seconds. Protein synthesis is tightly regulated. Translation consists of 3 major steps. 1. Chain initiation, 2. Chain elongation and 3. Chain termination. In addition to these, activation of amino acids prior to their incorporation into poly peptides and post translational processing of completed polypeptide constitute two additional stages.

Activation of amino acids: This process takes place in the cytoplasm. Each amino acid is attached to a specific tRNA and the reaction is catalyzed by amino acyl tRNA synthetase.

Amino acid + tRNA + ATP ⟶ Amino acyl - tRNA +AMP + ppi

This reaction occurs in two steps.

1.ATP +Amino acid ⟶ Amino acyl adenylic acid +pyrophosphate

2. Amino acyl adenylic acid + tRNA → Amino acyl-tRNA + adenylic acid

Generally methionine is first attached to tRNA and then formyl group is transferred finally forming fMet tRNA. (fMet = formyl methionine)

Initiation: Polypeptide synthesis begins from the amino terminal end. This stage requires 1) 30S and 50S sub units, 2) mRNA, 3) fMet tRNA, 4) a set of initiation factors IF1, 2 and 3 and 5) GTP (guanosine triphosphate) and Mg++. The formation of initiation complex is divided into three steps. In the first step, 30S subunit binds to IF (initiation factor) 3 which prevents the premature union of 30S and 50S. Then binding of mRNA to 30S sub unit takes place in such a way that the initiation codon AUG binds to a precise location on 30S. In the second step, complex of 30S, IF3 and mRNA form a still larger complex by binding IF2 (which is already bound to GTP) and the fMet tRNA. The anticodon of this tRNA pairs correctly with the initiation codon. In the third step this large initiation complex of 30S subunit combines with 50S subunit. Simultaneously GTP is hydrolyzed to GDP (guanosine diphospate) +Pi. IF2 and IF3 depart from ribosomes. This elaborate initiation process is required to ensure that the initiating aminoacyl- t-RNA is bound at the peptidyl site and positioned at the initiation codon AUG, so that the ribosomes start translation at the correct point on the mRNA. Translation of the codons of mRNA takes place in the 5'___3' direction.

Elongation: This stage requires the complex formed above, the next amino acyl tRNA, a set of elongation factors EF, Tu, Ts, G and GTP.

a) *Binding of incoming amino acid:* Elongation factor (EF)is made up of two subunits ie: Tu and Ts. As long as Tu and Ts are associated the next amino acid cannot bind to this complex. Once the two factors separate out then GTP will bind to Tu and forms a complex Tu-GTP and this complex will bind with the next amino acyl tRNA. This entire complex is bound to A site of ribosome and the energy released by GTP hydrolysis will facilitate the correct binding of the amino acid. GDP and EF are released.

b) *Formation of the peptide bond:* A new peptide bond is formed between the amino acids at A and P sites. This occurs by transferring formylmethionyl group from its tRNA to the amino group of second amino acid, which is at A site. The A site alpha amino acid acts as a nucleophile and formation of peptide bond takes place. The dipeptide thus formed is bound to A site leaving P site with uncharged tRNA at P site. Peptidyl transferase enzyme will participate in this reaction. For the formation of one peptide bond four ATP molecules are utilized.

c) Translocation: Finally the ribosome moves by a distance of one codon towards 3' end of mRNA, thus shifting dipeptidyl tRNA from A site to P site. Now A site is free for the incoming amino acyl tRNA. The uncharged tRNA is released from P site back into cytoplasm. The ribosome moves from codon to codon along mRNA towards 3' end, adding one amino acid at a time to the growing chain. Elongation continues until the ribosome adds the last amino acid, completing the polypeptide coded by the mRNA.

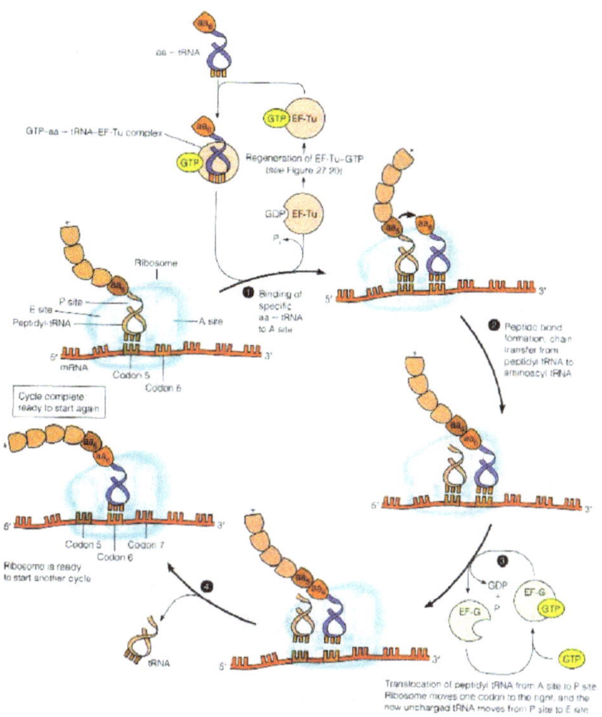

Fig. 2. Polypeptide biosynthesis in ribosomes

Termination: This stage is signalled by one of the 3 termination codons in mRNA – UAA, UAG, and UGA. The termination codon first occupies A site. Then the release factors 1, 2 and 3 contribute to the following: a) Hydrolysis of terminal peptidyl tRNA bond, b) Release of free polypeptide and last tRNA from P site, c) Dissociation of 70S subunit into 30S & 50S which are ready to start a new cycle of protein biosynthesis.

In bacteria RF1 recognizes the termination codons UAG and UAA, and RF2 reconizes UGA and UAA. RF1 or RF2 binds at a termination codon and induces peptidyl transferase to transfer the growing peptide chain. The specific function of RF3 has not been firmly established. In eukaryotes, a single release factor called eRF recognizes all three termination codons.

Inhibitors of protein biosynthesis: Protein synthesis being a long complex process is vulnerable to inhibition at many points. Some inhibitors of protein synthesis are

a) *Streptomycin:* inhibits initiation and causes misreading(binds to 30S subunit)

b) *Tetracycline:* in bacteria it inhibits by binding to amino acyl tRNA

c) *Chloramphenicol* : in bacteria it inhibits peptidyl transferase activity.

d) *Puromycin:* causes premature release of peptide, as it has a structure very similar to the 3' end of an aminoacyl-tRNA.

e) *Cycloheximide:* blocks the peptidyl transferase of 80S eukaryotic ribosome

f) *Ricin:* inactivates 60S subunit of eukaryotic ribosome

g) *Diphtheria toxin:* inactivates eukaryotic elongation factor eEF2

Post translational modifications: After synthesis, the nascent polypeptide chain is folded and processed into its biologically active form. It spontaneously assumes its native conformation i.e secondary, tertiary and quaternary structures. Some newly made proteins do not attain their final biologically active conformation until they have been altered by one or more processing reactions called post-translational modifications. The post translational changes can be grouped under the following sub-headings

1. **Amino terminal and Carboxyl terminal modification:** Protein synthesis is initiated by formyl-methonine or methionine which has to be removed enzymatically. The N terminal amino acid is acetylated.

2. **Loss of signal sequence:** 15 to 30 residues at the amino-terminal end of some proteins play a role in directing the protein to its ultimate destination in the cell. Such signal sequences are ultimately removed by specific peptidases.

3. **Modification of individual amino acids:** The hydroxyl groups of certain amino acids like serine, threonine and tyrosine residues of some proteins are enzymatically phosphorylated by ATP. Carboxyl groups of glutamic acid are methylated, lysine is methylated. Hydroxylation of Proline to hydroxy proline is carriedout.

4. **Attachment of carbohydrate side chain:** Carbohydrate side chain of glycoproteins is attached covalently during or after the synthesis of the polypeptide chain.

5. **Addition of isoprenyl groups:** A number of eukaryotic proteins are isoprenylated, a thioether bond is formed between the isoprenyl

groups and a cysteine residue of the protein. In some cases, the isoprenyl group serves to help anchor the protein in a membrane.

6. **Addition of prosthetic groups:** Many prokaryotic and eukaryotic proteins require covalently bound prosthetic groups for their activity. These are attached to the polypeptide chain after it leaves the ribosome. Ex. The attachment of heme group to the cytochrome c
7. **Proteolytic processing**: Some viral proteins and proteases are initially synthesized as larger inactive precursor proteins called as zymogens which have to be trimmed to produce finally active forms Ex: trypsin is synthesized as trypsinogen
8. **Formation of disulfide cross-links:** Proteins to be exported from eukaryotic cells, after undergoing spontaneous folding into their native conformations, are often covalently cross-linked by the formation of intra-chain and inter-chain disulfide bridges between cysteine residues. The cross links formed in this way help to protect the native conformation of the protein molecule from denaturation in an extracellular environment that can differ greatly from that inside the cell.

Degradation of proteins into amino acids

Proteins are hydrolyzed by proteases and the amino acids are released. These amino acids undergo catabolism in the following ways. a. Transamination b. Oxidarive deamination. c. Non- oxidative deamination and d. Decarboxylation

Transamination: This reaction involves the transfer of an amino group from a donor amino acid to an acceptor α-keto acid to yield the α-ketoacid of the donor amino acid and the amino acid of the original α-keto acid. The reaction is catalyzed by an enzyme called aminotransferase or transaminase that requires PLP as co-enzyme for its activity.

Oxidative Deamination: Amino group from amino acid is oxidatively removed in this process. Glutamate dehrogenase enzyme catalyses the reaction.

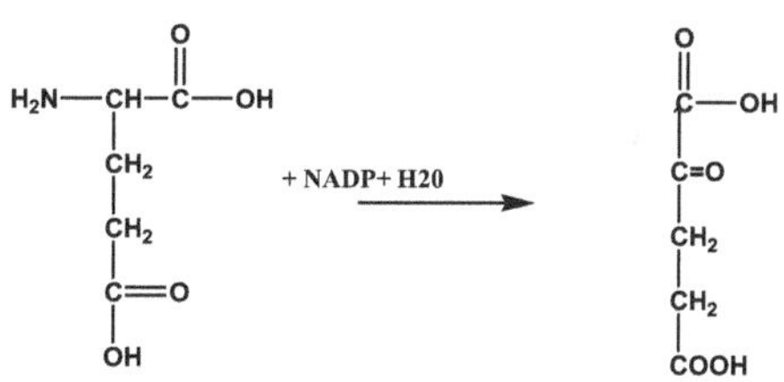

GLUTAMIC ACID ALPHA KETOGLUTARIC ACID

Non oxidative deamination: Non oxidative deamination takes place in absence of water and the corresponding unsaturated dicarboxylic acid is formed.

Aspartic acid + H2 ⟶ Fumaric acid + NH3

Decarboxylation: In decarboxylation reactions, the CO2 from the corresponding amino acids are removed and biologically active amines are formed.

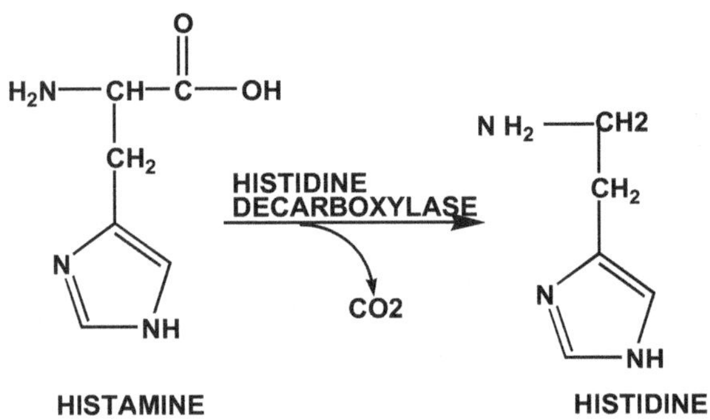

Amino acid carbon skeletons: The remainder of the amino acid is referred to as the "carbon skeleton". Depending on the particular amino acid being catabolised, its carbon skeleton will be converted to acetyl CoA, or pyruvate and or a citric acid cycle intermediate. Those carbon skeletons which end up as acetyl CoA are committed to energy production. They will be oxidised via the citric acid cycle.

Ammonia assimilation

The ammonia released during metabolism of amino acids is toxic to the cells if it is not incorporated into synthetic processes. There are three reactions that can catalyze the incorporation of the nitrogen atom as NH3 into organic compounds. These reactions are those catalysed by glutamic dehydrogenase, glutamine synthetase and carbamyl phosphate synthetase.

Details of amino acid catabolism are presented in the chapter on amino acid metabolism.

Chapter – 14

Terpenes

Primary metabolism in a plant comprises all metabolic pathways that are essential to the plant's survival. Primary metabolites, namely, carbohydrates, lipids and proteins are compounds that are directly involved in the growth and development of a plant whereas the secondary metabolites are compounds produced in other metabolic pathways that, although important, are not essential to the functioning of the plant. However, secondary plant metabolites are useful in the long term, often for defense purposes, and give plants characteristics such as color.

Secondary plant metabolites are used in signalling and regulation of primary metabolic pathways. Plant hormones, which are secondary metabolites, are often used to regulate the metabolic activity within cells. Secondary plant metabolites help the plant maintain an intricate balance with the environment, often adapting to match the environmental needs. Plant metabolites that color the plant are a good example of this, as the coloring of a plant can attract pollinators and also defend against attack by animals. Terpenes, alkaloids, sterols, pigments, growth regulators, and phenolics are the important secondary plant metabolites.

Terpenes are hydrocarbon based natural products, the structures of which are derived from isoprene units. Isoprene is a 5 C unsaturated hydrocarbon i.e. 2 – Methyl 1,3 – butadiene. Terpenes are thus referred to as Isoprenoids also. In simpler terms, terpenes are organic compounds containing an integral number of 5 C units. The name 'terpene' has now almost been replaced by the more accepted term 'terpenoids', since a large number of their oxidized products like alcohols, aldehydes, ketones etc. are derived from terpenes. Terpenes constitute one of the most diverse and numerous class of organic compounds. Also, they are very widespread and abundant too.

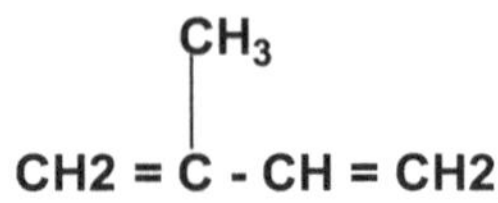

ISOPRENE

Classification

The different terpenes are classified based on the number of isoprene units they contain. Mono, di, and tri terpenes contain two, four and six isoprene units while the poly terpenenes contain many isoprene units. Various terpenes based on the number of isoprene units are presented in the Table 1.

Table 1. Terpenoids with varying isoprene units in plants

S.No.	No. of C atoms	No. of 5C units	Class	Examples
1.	5	1	Hemiterpenes	Iso amyl alcohol, isovaleric acid
2.	10	2	Monoterpenes	Menthol (peppermint), geraniol (rose oil), limonene (orange, lime), citronellal (citronella oil)
3.	15	3	Sesquiterpenes	Abscissic acid(plant growth regulator), zingeberene (ginger), farnesol (lily)
4.	20	4	Diterpenes	Gibberellins (plant growth regulator), gossypol (cotton seed, toxic), phytol (chlorophyll), retinol (role in vision)
5.	25	5	Sesterpenes	Very rare in nature
6.	30	6	Triterpenes	Squalene (shark liver oil), cholesterol, lanosterol (both precursors of hormones)
7.	40	8	Tetraterpenes	β-carotene (carrot, tomato), lycopene (tomato)
8.	Many	Many	Polyterpenes	Rubber, gutta

Functions of terpenes in plants

1. As plant growth regulators: Important among them are abscisic acid, gibberellins and brassinolide, a sterol.
2. As electron carriers: Ex. Ubiquinone in electron transport system.

3. As photosynthetic pigments: Ex. β – Carotene
4. As constituents of cell membranes: Ex. Steroids.
5. Plant defense/ protection agents: Ex. Damaged plants exude resinous materials. The role of such resins (mostly terpenes) is to seal the wound, so that the infecting and attacking microbes fail to enter inside. Arbutin, a terpenoid, is an allellopathic chemical inhibiting the growth and germination of other plants nearby. Even, the major insecticidal principles of neem are terpenoids Ex. Azadirachtin.
6. Insect repellents: Ex. Citronellal in citronella oil.
7. Pollination helpers: Ex. Perfumes present in many flowers help attract insects and result in pollination.
8. Fungal antibiotics Ex. β–Phellandrenes.

Uses of terpenes

1. Natural rubber (a polyterpene - major source- *Hevea brasiliensis*) is used in the manufacture of vulcanized and high strength, less friction rubber.
2. As flavoring agents in foods, drinks, ice creams and even cigarettes. A widely used such material is Vanillin present in vanilla beans.
3. Perfumer Ex. Geraniol is widely used as a perfumery material.
4. Oil of turpentine is used as a solvent in paints.
5. As fragrances in soaps, talcum powders and cosmetics.
6. Insect repellents Ex. oil of citronella.
7. Insect control agents Ex. as juvenile hormones which prevent metamorphosis in insects and neem based insecticidal preparations.
8. In medicine: Ex. a) Santonin has been used in India since long for the control of intestinal worms. b) Digitoxin is used in heart beat regularization. c) Cineole is used as a nasal decongestant.

Chapter – 15

Phenolics

Phenol is normally a benzene ring containing an –OH group, and hence an aromatic alcohol. The term phenolic refers to similar aromatic compounds with one or more -OH groups. Phenolics constitute one of the most widespread classes of secondary metabolites. Commonly talked about class of compounds like anthocyanins, flavonoids, lignins, tannins, coumarins and chalcones are all phenolic substances. Phenolics are water insoluble substances. However, the water solubility increases with the increase in the number of -OH groups. Phenolics absorb electromagnetic radiation strongly. Many of the phenolics absorb in the visible part of the spectrum (Ex. anthocyanins) and are coloured. Phenolics are produced in plants by 1. Shikimate pathway through phenylalanine (C6.C3) and 2. Acetate/ malonate pathway.

Some important phenolic compounds

Phenolic compounds occur exclusively in the form of glycosides and esters. Polyphenols like tannins, coumarins, flavonols, anthocyanins and hydroxylated cyclohexanes are present in plants. Chlorogenic acid, caffeic acid, ferulic acid, quinic acid, scopoletin, rutin, coumaric acid, eugenol, guaicol, vanillic acid etc. are some important phenolics in plants.

Lignin, a cell wall strengthening material, is a C6-C3 polymer. It is more abundant in wood, corky tissues and straw. Lignin imparts resistance to degradation by water and microorganisms. In plant cell walls where lignin is more common, lignin remains embedded in cellulose microfibrils in a dense cross connected network. Lignin is made up of three aromatic alcohols of which the most important is p-coumaryl alcohol.

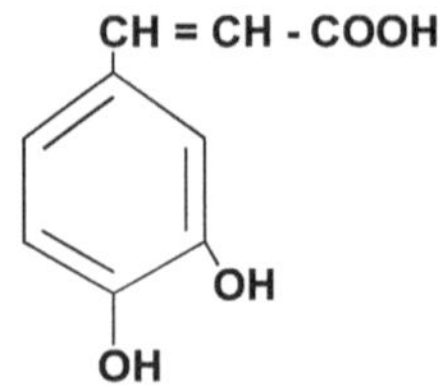

CAFFEIC ACID

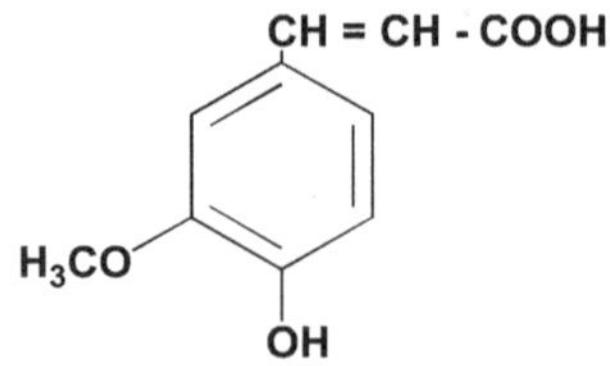

FERULIC ACID

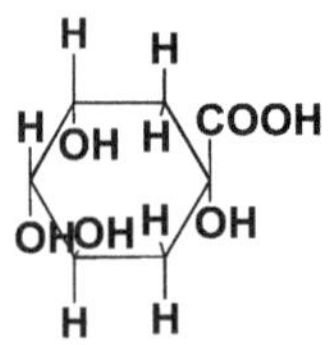

QUINIC ACID

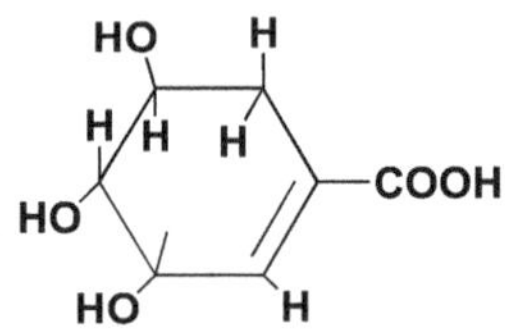

SHIKIMIC ACID

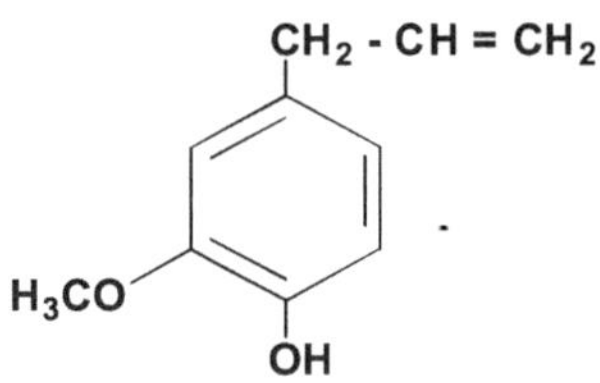

EUGENOL

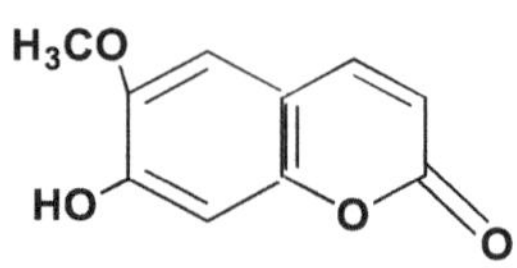

SCOPOLETIN

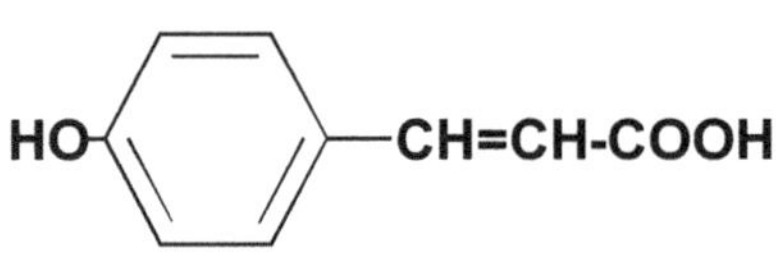

p- COUMARIC ACID

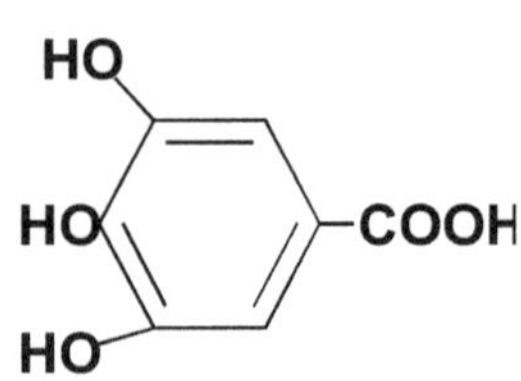

GALLIC ACID

Fig. 1. Some phenolic compounds

Flavonoids are 15 carbon phenolics with C6.C3.C6 carbon skeleton. Flavonoids are widely distributed in the plant kingdom. They absorb radiation very strongly in the visible part of the electromagnetic spectrum and are differently coloured. Anthocyanins which impart colour to flowers, leaves, and fruits of many plants are flavonoids. In flowers, they help attract the pollinating insects. A specific example of flavonoid is delphinidin, the substance that gives blue color to grapes. Flavonols and flavones are also important flavonoids. Flavonoids occur as glycosides attached to sugars, normally glucose and occasionally galactose.

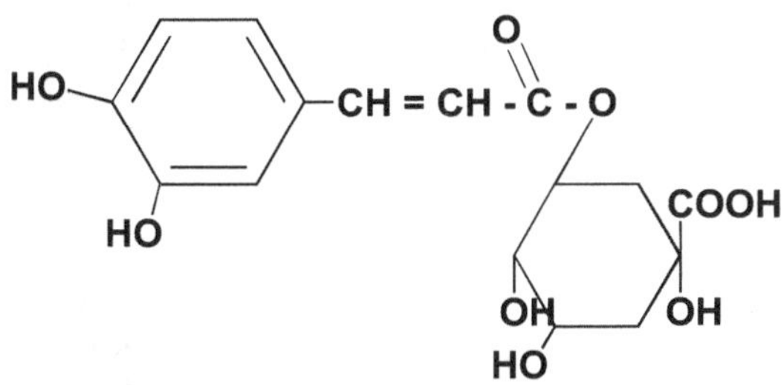

CHLOROGENIC ACID (3-O-CAFFEOYL-d-QUINIC ACID)

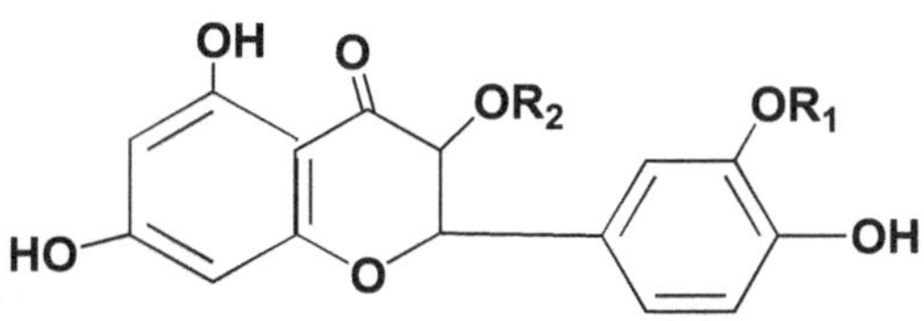

RUTIN = R1=OH, R2 = l- RHAMNOSIDO
d - GLUCOSYL

FLAVONOID

Tannins are C6.C3.C6 ploymeric phenolics. They are astringent phenols and tan animal skin to leather. Tannins cross link proteins, thus denaturing them. They thus keep the microorganisms away and impart a protective function. Coumarins form the starting metabolite of a number of important phenolics. Aroma of freshly cut grass is due to the coumarins.

Classification of phenolics

The phenolics are classified based on their carbon skeletons. Different phenolics along with examples are presented in the Table 1. The classification is mostly based on the number of carbons and different group combinations. Various phenolics ranging from simple to very complex phenols occur in plants.

Table 1. Different phenolics, their classification and examples

S no	No of CAtoms	C skeleton	Class	Examples
1	6	C-6	Simple phenols	Hydroquinone, catechol
2	7	C6-C3	Hydroxyl benzoates	4-Hydroxybenzoate
3	8	C6.	C2 Phenylacetates	Hydroxyphenyl acetate
4	9	C6-C3	1. Hydroxy cinnamates 2. Phenylpropenes 3. Coumarins 4. Chromones	1. Caffeate 2. Eugenol 3. Esculetin 4. Eugenin
5	10	C6.C4	Naphthoquinones	Juglone
6	13	C6.C1.C6	Xanthones	Mangiferin
7	14	C6.C2.C6	Stilbenes	Resveratrol (eucalyptus)
8	15	C6.C3.C6	Flavonoids	Rutin
9	18	(C6.C3)2	Lignans	Pinoresinol
10	30	(C6.C3.C6)2	Biflavonoids	Amentoflavone
11	Many	(C6)n	Melanins	———
12	Many	(C6.C3)n	Lignins	———
13	Many	(C6.C3.C6)n	Condensed tannins	———

Functions of phenolics

1. **Pigmenatation:** Pigmentation of the outer surface and inner parts of plants and animals is one of the important functions of phenolics. Thus, melanin is responsible for the darkening of the skin in human beings. Chlorogenic acid imparts darkening of the potato tuber. Browning of the cut tissues of many fruits is also due to the presence of phenolic substances.

2. **Regulation of plant growth:** Important phenolics regulating plant growth are ferulic acid and phytoalexins.

3. **Disease resistence:** Protocatechuic acid imparts resistance to smudge caused by *Colletotrichum circinam* in onions.

4. **Alleollopathy:** Gallotannin obtained from gallic acid is an allellochemical in many plants.

5. **Insect control:** A number of coumarins act as juvenile hormones in insects by preventing metamorphosis and hold promise in insect control.
6. **Electron carrier:** Some phenolics act as electron carriers Ex. ubiquinone /ubiquinol in Electron Transport System.
7. **U. V. protection:** Phenolics help protect animals / plants against damage by Ultra Violet radiation.
8. Phenolics play an important role in leaf colour, quality and physiological strength of smoke in the case of tobacco.

Industrial uses of phenolics

1. Manufacture of medicines: Aspirin, a widely used chemical as antipyretic and heart beat regulator is manufactured from simple phenolic compounds.
2. Phenolics (polymers) are used for making adhesive resins.
3. Phenolics (polymers) are used as electrical insulating materials.
4. Phenolics are used in the manufacture of pesticides like 2:4-D (weedicide), and carbaryl (insecticide).
5. Phenolics are widely used ingredients in cosmetics, particularly sun creams.
6. Phenolics are used in the manufacture of plastics like Bakelite.
7. Phenolics are extensively used in the making of coloring materials.

Cyanogenic glycosides

Many plants have adapted to iodine-deficient terrestrial environment by removing iodine from their metabolism. The cyanogenic glycosides are found in over 2000 plant species. Many plant pesticides are cyanogenic glycosides which liberate cyanide. The cyanide is poisonous to parasites and herbivores. The cyanogenic glycoside is useful in seed dormancy phase. Cyanogenic glycosides stored in the seed of the plants are metabolized during germination to release nitrogen for the seedling to grow.

Phytic Acid

Phytic acid is the main method of phosphorus storage in plant. It also is a source of energy and cations, a natural antioxidant for plants, and can be a source of myoinositol which is one of the preliminary pieces for cell walls.

Phytic acid is also known to bond with many different minerals, and by doing so prevents those minerals from being absorbed; making phytic acid an anti-nutrient. There is a lot of concern with phytic acids in nuts and seeds because of its anti-nutrient characteristics. Cooking can reduce the amount of phytic acid in food but soaking is much more effective. Phytic acid is an antioxidant found in plant cells and serves the purpose of preservation. This preservation is removed when soaked, reducing the phytic acid and allowing the germination and growth of the seed. When added to foods it can help prevent discoloration by inhibiting lipid peroxidation. There is also some belief that the chelating property of phytic acid may have potential use in the treatment of cancer.

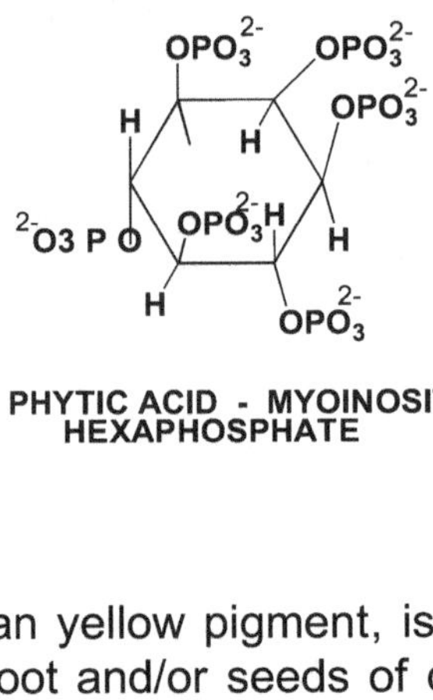

PHYTIC ACID - MYOINOSITOL HEXAPHOSPHATE

Gossypol

Gossypol ($C_{30}H_{30}O_8$), an yellow pigment, is found in cotton plants. It occurs mainly in the root and/or seeds of different species of cotton plants. Gossypol exists in three forms: gossypol, gossypol acetic acid, and gossypol formic acid. All of them have similar biological properties. Gossypol is a polyphenolic aldehyde, and has a formyl group. The formation of gossypol occurs through an isoprenoid pathway. Gossypol's main function in the cotton plant is to act as an enzyme inhibitor. An example of gossypol's enzyme inhibition is its ability to inhibit nicotinamide adenine dinucleotide-linked enzymes of *Trypanosoma cruzi. Trypanosoma cruzi* is a parasite which causes Chaga's disease. It is an effective protein kinase C inhibitor. It also inhibits replication of HIV-1 virus. It is an inhibitor of several dehydrogenase enzymes.

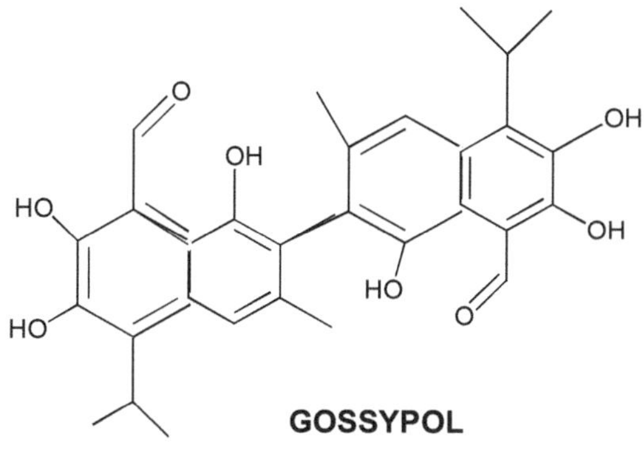

GOSSYPOL

Gossypol has been considered as a waste product produced during the processing of cottonseed. Recent studies have shown that gossypol has other functions. It can act as a male contraceptive. Gossypol has also been linked to causing hypokalemic paralysis, a disease characterized by muscle weakness. Gossypol induced hypokalemic paralysis is easily treatable with potassium repletion. Cotton seeds contain gossypol in the range of 0.7-1.2 %.

Phytoestrogens

Plants synthesize certain compounds called secondary metabolites which are not naturally produced by the humans but can play a vital role in protection or destruction of human health. One such group of metabolites is phytoestrogens. They are chemicals which act like the hormone estrogen. Phytoestrogens are found in oilseeds like groundnut and soybean . Estrogen is important for women's bone and heart health, but high amounts of it has been linked to breast cancer. In the plant, the phytoestrogens are involved in the defense system against fungi. One role of estrogens which phytoestrogens mimic is its protective behavior for the heart. So, an intake of phytoestrogens reduces the risk of cardiovascular disease. Resveratrol, a phytoestrogen found in grapes and groundnut is responsible for such a role.

Chapter – 16

Plant Pigments

Plant pigments impart colour to the different components like leaves, flowers, fruits etc. They in turn are responsible for the colour of raw food, we eat. Pigments like chlorophyll, anthocyanins, flavonoids, tannins, xanthones etc. are the different pigments that are present in plants. The primary function of pigments in plants is photosynthesis, which uses the green pigment chlorophyll along with several red and yellow pigments that help to capture as much light energy as possible. Other functions of pigments in plants include attracting insects to flowers to encourage pollination.

Table 1. Different plant pigments

Class	Compounds	Colour imparted
Porphyrins	Chlorophyll	Green
Carotenoids	α, β and γ - carotenoids and lycopene	Yellow
Flavonoids	Flavones	Yellow
	Flavonol	Yellow
	Anthocyanin	Purple, Magenta

Chlorophyll

Chlorophylls are green pigments involved in photosynthesis in plants and some bacteria. Chlorophyll is the primary pigment in plants. It absorbs yellow and blue wavelengths of light while reflecting green. That is why chlorophyll gives plants their green color. All chlorophylls serve as the primary means to intercept light to induce photosynthesis. Chlorophylls are complex molecules containing four pyrrole rings with 'Mg' atom in the centre. This is in contrast to the heme molecule, the constituent of blood in animals, where the nitrogen atoms are coordinated with the 'Fe' atom to form an iron porphyrin. The conjugate tetra pyrroles are known as porphyrins.

There are a number of chlorophylls, namely, chlorophyll a, b, c, and d. They are present in the green parts of the plants. Mainly chlorophylls a and b are present in the green plants especially the leaves. They differ slightly in their structures. Chlophyll a has a methyl group in ring iv while chlorophyll b has a -CHO in place of methyl group. In the same ring both the chlorophylls have a phytol group esterified to the propinonic acid of ring iv. Chlorophylls a and b are present in 1:3 ratio. They differ in the wave length of light they absorb. Chlorophyll c has only a -CH=CH COOH in place of the propionic and the phytyl groups. Chlorophyll d has a CHO group in ring I in place of CH=CH2.

The chlorophylls have characteristic absorption and fluorescence spectra, which are useful in the analysis of the pigments. In the presence of an acid the central Mg of the chlorophyll is replaced by hydrogen giving pheophytins, with a change in colour from green to dull olive brown. The phytol group is removed by enzyme chlorophyllase resulting in the formation of chlorophyllides which are green compounds.

When chlorophyll is subjected to sunlight some of the electrons get excited, raising the energy level of the molecule. Chlorophyll a is concerned in the production of NADPH + H^+ with the help of radiation less than 700nm. Chlorophyll b produces the oxidant leading to the formation of oxygen using 680nm light. The overall reaction is

$$\mathbf{2NADP+ + 2H2O \longrightarrow 2NADPH +2H+ + O2}$$

The interaction between the weak reductant and weak oxidant results in the flow of electrons followed by generation of ATP. This is called photophosphorylation. Photosynthetic process converts 10^{11} tonnes of CO_2 to carbohydrates annually. This is equivalent to 10^{17} k calories of free energy storage. The chlorophyll of the plants is the main architect of this process.

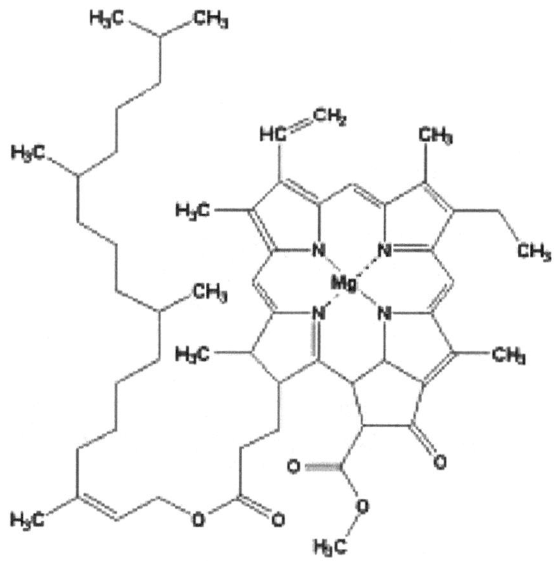

Chlorophyll A

Carotenoids

Carotenoids are a group of lipid soluble hydrocarbons and their oxygenated derivatives (xanthophylls). They derive their name from the principal colouring matter of carrot roots, namely, carotene. They are widely distributed in the plant world. Carotenoids are organic pigments present in the chloroplasts and chromoplasts of plants. There are over 600 known carotenoids. In plants, carotenoids occur in roots, stems, leaves, flowers, and fruits.

Carotenoids consist of eight isoprenoid units joined in such a way that the isoprene units are reversed at the centre of the molecule. The carotenoids found in nature may be regarded as derivatives of the red pigment lycopene. Lycopene has the empirical formula $C_{40}H_{56}$. It is a highly unsaturated straight chain hydrocarbon (polyene) composed of two identical units joined by a double bond. Each of the $C_{20}H_{28}$ unit is made of four isoprene (CH_2=C(CH_3)-CH=CH_2) units. Carotenoids are red, orange, or yellow tetraterpenoids. They function as accessory pigments in plants, helping to fuel photosynthesis by gathering wavelengths of light not readily absorbed by chlorophyll. The most familiar carotenoids are carotene (an orange pigment found in carrots), lutein (a yellow pigment found in fruits and vegetables), and lycopene (the pigment responsible for the red color of tomatoes).

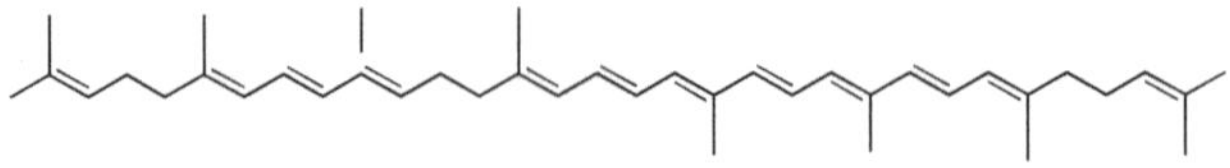

LYCOPENE

The basic carbon structure for lycopene indicates the symmetrical arrangement around the central pair of carbons. This structure cyclized at the ends is beta carotene. Carotenoids exist in nature in the free state. In plant tissues they exist as crystalline or amorphous solids or in solution in lipid media. The alternate double bonds are responsible for their colour. The trans configuration offers a stable structure and hence carotenoids exist in their trans forms. The orange red color of the trans compounds change to lemon yellow when they change to cis configuration. Acids and temperature facilitate this change.

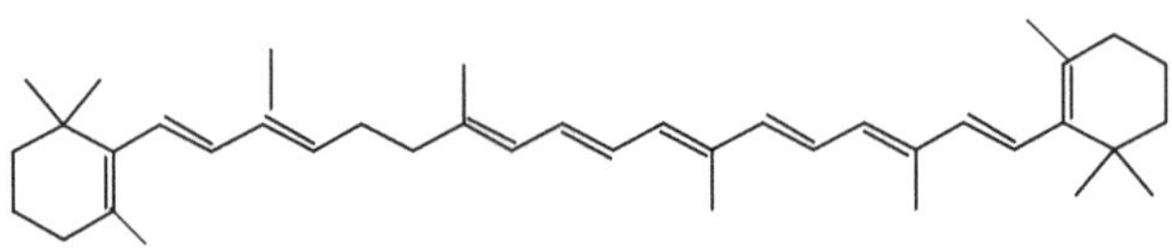

β carotene

Three carotenes, namely, α, β and γ carotenes exist in nature. A characteristic feature of the carotenes is the presence of a ring at one or both ends of the hydrocarbon chain. Alpha and beta carotenes have rings at both ends, but differ only with respect to a double bond in one of the rings. Gamma carotene has only one ring at one of the ends of the chain. This ring is structurally related to the compounds termed alpha and beta ionones. Ring one of alpha, beta and gamma – carotenes and ring two of beta carotene correspond to a beta ionone residue, while ring two of alpha carotene correspond to an alpha ionone residue.

In green plants carotenoids occur together with chlorophyll in the chloroplasts. The phytol residue of the chlorophylls may be formed from Beta carotene is a precursor of vitamin A. Carotenoids have been shown to act as antioxidants and to promote healthy eyesight in humans. Carotene yields two molecules of vitamin A by cleavage at the centre. Vitamin A (C20H29-OH) is essentially an oxidation product of one half of the beta carotene structure. Beta carotene is converted in the animal systems to vitamin A. It is found in large amounts in the liver of sal fish.

Carotenoids are split into two classes, xanthophylls and carotenes. Xanthophylls are carotenoids with molecules containing oxygen, such as lutein and zeaxanthin. Carotenes are carotenoids with molecules that are unoxygenated, such as α-carotene, β-carotene and lycopene. Carotenoids have two important functions in plants. First, they can contribute to photosynthesis. They do this by transferring some of the light energy they absorb to chlorophylls, which then uses this energy for photosynthesis. Second, they can protect plants which are over-exposed to sunlight. They do this by harmlessly dissipating excess light energy which they absorb as heat. In the absence of carotenoids, this excess light energy could destroy proteins, membranes, and other molecules. Also they may function as regulators of certain developmental responses in plants. Carotenoids involved in photosynthesis are formed in chloroplasts. Others are formed in plastids.

Vitamin A_1 is an oxidation product of one half of the β carotene molecule. β carotene is convered in the animal systems to vitamin A_1 and is found in the liver of fish etc. Vitamin A_2 contains an extra double bond over vitamin A_1. Carotenoids have been considered important due to their relationship to vitamin A activity. In green plants carotenoids occur together with chlorophyll in the chloroplasts. The phytol residue of chlorophylls is considered to be derived from the carotenoids.

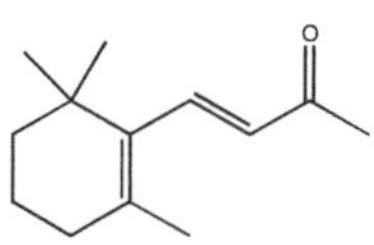

BETA IONONE

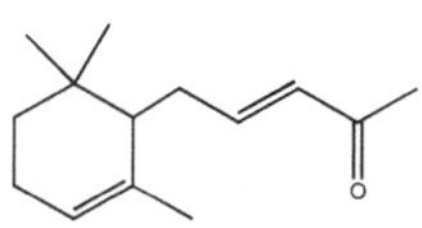

ALPHA IONONE

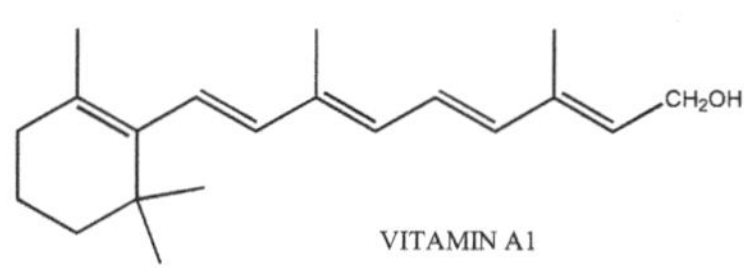

VITAMIN A1

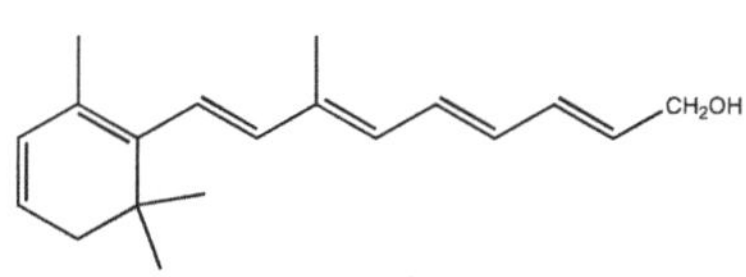

VITAMIN A2

Anthocyanins and flavonoids

Anthocyanins (literally “flower blue”) are red, blue and violet pigments. They are glycosides containing one or two carbohydrate units and an aglycone known as anthocyanidin. The anthocyanidins are derivatives of 3,5,7 –trihydroxy flavylium hydroxide (2-phenyl, 3,5,7 trihydroxy benzo pyrilium hydroxide). These pigments are present in flowers and other coloured plant parts.

Three main groups of anthocyanidins are identified based on the substitution in ring B. The colour of the pigments depend on the number of hydroxyl groups and or the methoxyl groups. The colour of the solution of an anthocyanidin also depends on the p^H. The pigments are generally red in acid solution, violet or purple in neutral and blue in alkaline solutions.

Another widely distributed group of water soluble plant pigments related structurally to the anthocyanidins are derivatives of 2-phenyl 1-4 benzopyrone (flavones). Flavones have been isolated from the primrose and other plants. Most of their hydroxylated derivatives are classified as flavones, flavonols and flavanones. Higher plants may contain two other groups of yellow pigments, namely, chalcones, which are derivatives of benzalacetophenone and aurones, which are benzal coumaran-3-one. These pigments occur in their free as well as their glycosidic forms.

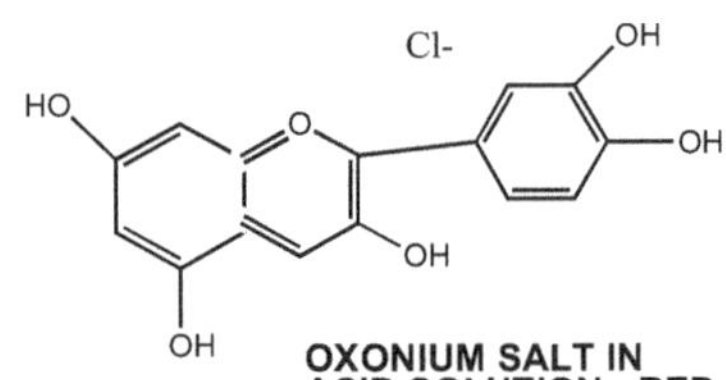

OXONIUM SALT IN ACID SOLUTION - RED

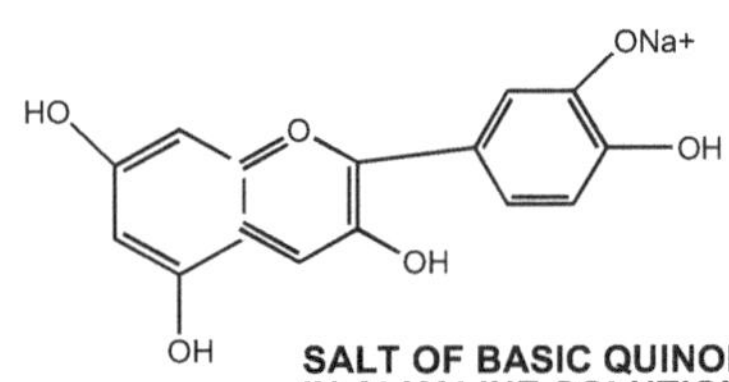

SALT OF BASIC QUINONE IN ALKALINE SOLUTION BLUE

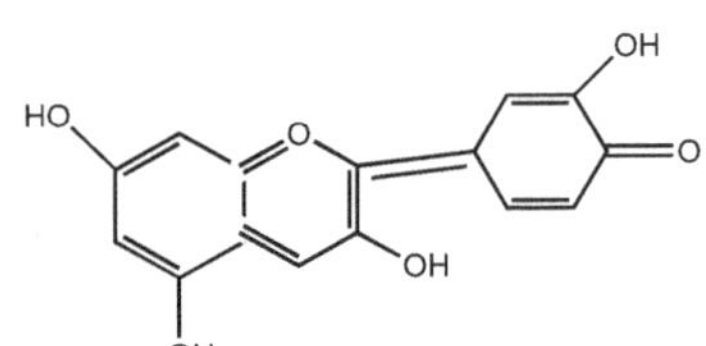

FREE BASE IN NEUTRAL SOLUTION- VIOLET

Flavonoids are a class of secondary plant metabolites. They are also known as Vitamin P or citrin. These metabolites are mostly used in plants to produce yellow and other pigments which play a big role in coloring the plants. Flavonoids have anti-inflammatory, anti-allergic and anti-cancer activities. Flavonoids also have anti-oxidant properties. Flavonoids like hesperidins, quercitrin,and rutin, are present in berries, tea, dark chocolate and red wine. Many of the health benefits attributed to these foods come from the flavonoids they posses. Flavonoids are closely related to flavones which are actually a sub class of flavonoids, and are the yellow pigments in plants. In addition to flavones, there are 11 other subclasses of flavonoids including, isoflavones, flavans, flavanones, flavanols, flavanolols, anthocyanidins, catechins (including proanthocyanidins), leukoanthocyanidins, dihydrochalcones, and aurones.

These pigments are variants of the C_6- C_3-C_6 structure. The fact that several species of plants contain esters of p-coumaric and caffeic acids suggest that these C_3-C_6 acids have a biosynthetic relation to the corresponding C_3-C_6 (ring B) portions of aglycones. Probably a phenyl propane derivative condenses with polyphenol to yield C_6- C_3-C_6 structure.

Anthocyanins, the water-soluble flavonoid pigments occur in all tissues of higher plants, providing colour in leaves, plant stem, roots, flowers, and fruits, in small quantities. Anthocyanins are most visible in the petals of flowers, where they account for as much as 30% of the dry weight of the tissue.

ANTHOCYANIDINDS

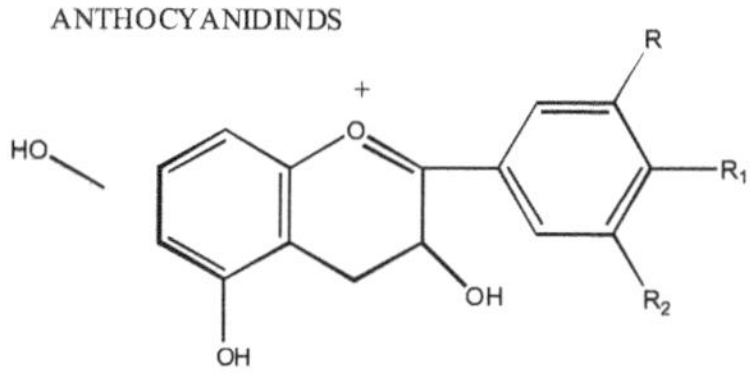

ANTOCYANINS	R	R1	R2
PELARGONIDIN	H	OH	OH
CYANIDIN	OH	OH	H
DELPHINIDIN	OH	OH	OH
PEONIDIN	OCH3	OH	H
MELVIDIN	OCH3	OH	OCH3

FLAVONES AND FLAVONOLS

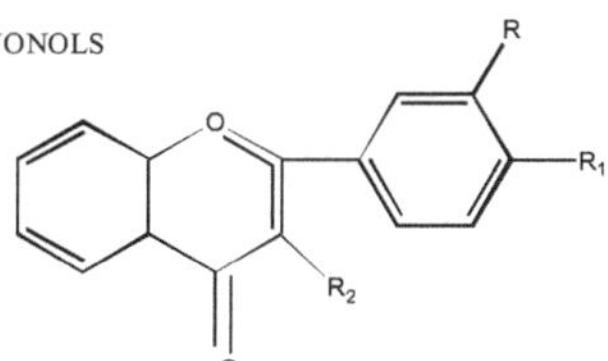

	R	R1	R2
CHRYSIN	H	H	H
APIGENIN	H	OH	H
LUTEOLIN	OH	OH	H
KAEMPFEROL	H	OH	OH
QUERCETIN	OH	OH	OH

Betalains are red or yellow pigments. Like anthocyanins they are water-soluble, but unlike anthocyanins they are indole-derived compounds synthesized from tyrosine. This class of pigments is found only in the *Caryophyllales* (including cactus and amaranth), and never co-occur in plants with anthocyanins. Betalains are responsible for the deep red color of beets, and are used commercially as food-coloring agents.

Chapter – 17

Sterols

Sterols are derivatives of a complex ring system called cyclopentano perhydro phenanthrene. The term sterol means solid alcohol. Sterols, though alcohols, are solids with a M.P. of 100-120^0 C. The 27 carbon atoms impart hydrocarbon character, namely, solubility in non-polar solvents. Free sterols in plants are mono hydroxy secondary alcohols varying in the constitution of methyl or ethyl group and in double bonds. Cholesterol, ergosterol, bile acids, sex harmones, cortical harmones and D-vitamins are some of the sterol derivatives.

The most widely distributed sterols in higher plants are the C_{29} sterols, stigmasterol, β sitosterol and γ sitosterol. Together, they are called steroids. In plants, steroids are present as glycosides. Cholesterol was first isolated from gall stones. The word cholesterol means "solid alcohol from bile". Bile is a fluid which is formed and stored in the gall bladder. It aids in the digestion of fats in the digestive tract. Sterols form an important group among the steroids. Unsaturated steroids with most of the skeleton of cholestane containing a 3β-hydroxyl group and an aliphatic side chain of 8 or more carbon atoms attached to position 17 form the group of sterols.

Phytosterols

The first plant sterol isolated was stigmasterol by Windaus and Hault in 1906. The term "phytosterol" was proposed in 1897 for all the sterols of vegetable origin. Chemically, these sterols have the same basic structure as cholesterol but differences arise from the lateral chain which is modified by the addition of one or two supernumerary carbon atoms at C-24 with either a or b chirality. The 24-alkyl group is characteristic of all phytosterols and is preserved during subsequent steroid metabolism in both fungi and plants to give hormones that regulate growth and reproduction in a manner similar to animals.

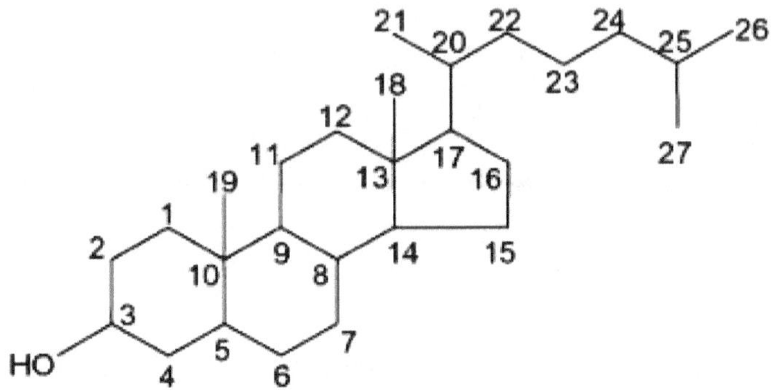

Fig. 1. Structure of sterols with carbon numbers

Phytosterols are more efficient than cholesterol in extending the temperature range in which membrane-associated biological processes can take place. This is because plants face higher temperature variations than animals. Most phytosterols are compounds having 28 to 30 carbon atoms and one or two carbon-carbon double bonds, typically one in the sterol nucleus and sometimes a second one in the alkyl side chain. All phytosterols have been shown to be derived in plants from cycloartenol.

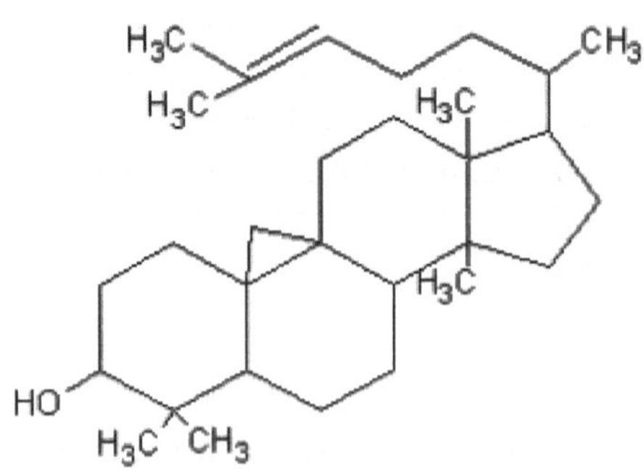

Cycloartenol

More than 250 different types of phytosterols have been reported in plant species. Representatives of these sterols are campesterol, β-sitosterol and stigmasterol (in vegetable oils). β-Sitosterol is present in all plant lipids and is used for steroid synthesis. Stigmasterol, which is used for the synthesis of progesterone and vitamin D3, is a potential anti-inflammatory compound. Its action is mediated by the inhibition of several pro-inflammatory and matrix degradation mediators involved in osteoarthritis-induced cartilage degradation.

An important sterol from yeast and ergot is the C28 compound ergosterol (mycosterol). Later it has been observed that fungi in general, contain ergosterol. Upon irradiation, this sterol gives rise to vitamin D2 (calciferol). As ergosterol is a cell membrane component largely restricted to fungi, its amount in environmental matrices may be used as an index molecule for these micro-organisms in a living biomass. Sterols are isolated from the unsaponifiable fraction of most lipid extracts and determined by various chromatographic procedures (HPLC or GLC).

Considerable variability in the concentration of free sterols has been observed among different oils. While concentrations lower than 100 mg/100 g are found in oils from coconut, palm, olive, and avocado, concentrations between 100 and 200 mg/100 g are found in oils from peanut, safflower, soybean, borage, cottonseed, and sunflower, and concentrations between 200 and 400 mg/100 g are found in oils from sesame, canola, rapeseed, corn and evening primrose.

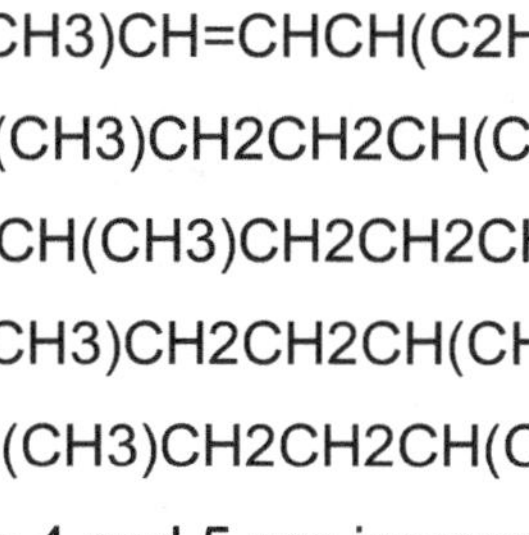

1. Stigmasterol R = CH(CH3)CH=CHCH(C2H5)CH(CH3)2
2. Betasitosterol R = CH(CH3)CH2CH2CH(C2H5)CH(CH3)2
3. Gammasitosterol R = CH(CH3)CH2CH2CH(C2H5)CH(CH3)2
4. Campesterol R = CH(CH3)CH2CH2CH(CH3)CH(CH3)2
5. Brassicasterol R = CH(CH3)CH2CH2CH(CH3)CH(CH3)2

2 and 3 are isomers, also 4 and 5 are isomers

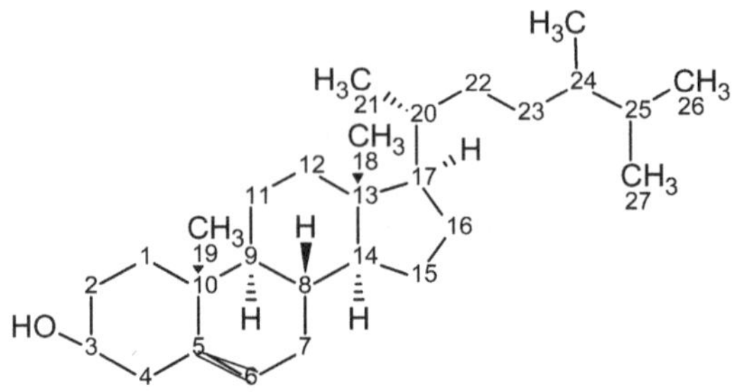

Campesterol

Phytostanols are a fully-saturated subgroup of phytosterols (they contain no double bonds). They are in general produced by the hydrogenation of phytosterols. These saturated sterols are found in very small amounts in plant products, such as nuts, seeds and legumes but in higher amounts in tissues of a few cereal species. To improve their solubility, plant stanols are often combined with a fatty acid ester to produce plant stanol esters. The most generally found stanol is sitostanol.

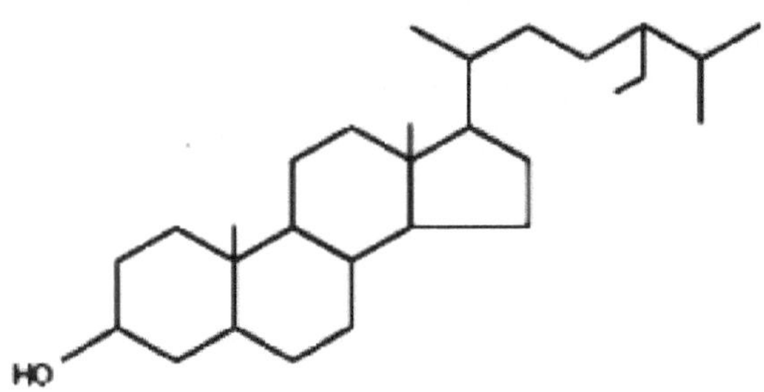

Sitostanol

Sterols in plant cells have physiological functions associated with cell membranes. Cholesterol, stigmasterol and β-sitosterol decrease the methanol initiated betacyanin leakage. Similar to their role in animal systems, sterols in plants also function as membrane stabilizers, as hormones and precursors of other steroids.

Steryl glycosides: Glycosylated steryl derivatives are synthesized by most plants and fungi, some animals and a few bacteria. This family consists of one carbohydrate unit linked to the hydroxyl group of one sterol molecule. The sterol moiety can be one of the sterols like campesterol, stigmasterol, sitosterol, brassicasterol and dihydrositosterol. The sugar moiety is composed of glucose, xylose and even arabinose.

β-sitosterol glucosides

R = H or $C_{15}H_{31}CO$

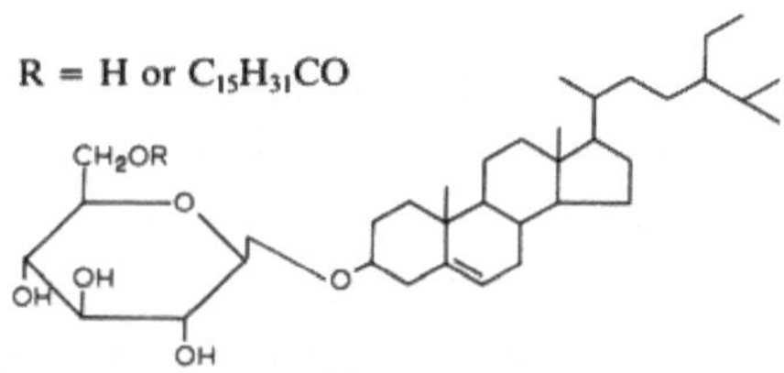

It has been observed that sterol glucosides participate in the synthesis of cellulose. This glycolipid is used as a substrate to produce higher homologues of the cellobioside type with β-1,4-linked glucosyl residues. The resulting disaccharide is split off and used as primer for further elongation to cellulose.

Sterol esters: Sterols occur as esters of palmitic, oleic, linoleic and linolenic acids in plant tissues. In plants, several sterol esters can be found in cell membranes and seed oils, such as ergosteryl, stigmasteryl and β-sitosteryl esters. Even cycloartenol and stigmasterol esters have been isolated. The relative importance of esterified sterols depends on the vegetable oil, 50-70% being found in oils from evening primrose, avocado, rapeseed, canola, corn, peanut, and sunflower, 30-50% in oils from borage, olive, sesame, coconut, and cottonseed, and less than 30% in oils from safflower, palm, and soybean. Thus, a large variation in the content and distribution of the sterol fractions between different vegetable oils can be observed.

In addition to variations in quantities, yeast sterol esters have been found to vary in both the sterol and fatty acid components. Fatty acids have a carbon chain from 12 up to 18 carbon atoms, saturated or unsaturated having one to three double bonds. More than 20 different sterols have been observed to occur in the esterified form. Their distribution as free sterols (S), steryl esters (SE), steryl glycosides (SG) and steryl acyl glycosides (ASG) in different plants is presented in the Table.

Table 1. Distribution of sterol derivatives in plant tissues (%):

Plant tissue	S	SE	SG	ASG
Potato	6.5	5.0	12.2	72.2
Apple fruit	68.8	7.2	18.5	5.5
Soyabean	59.0	15.0	11.0	15.0
Wheat endosperm	13.8	82.0	1.6	2.5
Tobacco leaf	64.3	13.5	18.0	4.0

Cholesterol, a C27 alcohol, is found in high concentrations in animal cell membranes. Concentrations ranging from 30 to 50 mol% (of total lipids) in red blood cells and as high as 80 mol% in the ocular lens membranes have been reported. Cholesterol has various functions in membranes ranging from metabolism, as a precursor to hormones and vitamins, to providing mechanical strength and a control of the phase behavior of membranes. In the tissues of vertebrates, cholesterol (Greek, *chole*, bile), is abundant in adrenals (10%, w/w), nervous tissues (2%,w/w), liver (0.2%,w/w) and gall stones. The vertebrate brain is the most cholesterol-rich organ, containing roughly 25% of the total free cholesterol present in the whole body.

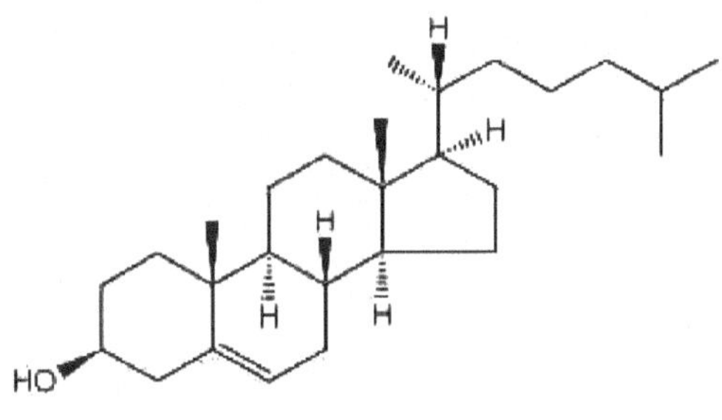

Cholesterol

Avenasterol can be isolated from oat oil. This sterol has been shown to protect specifically frying oils from oxidation owing to its ethylidene group in the side chain

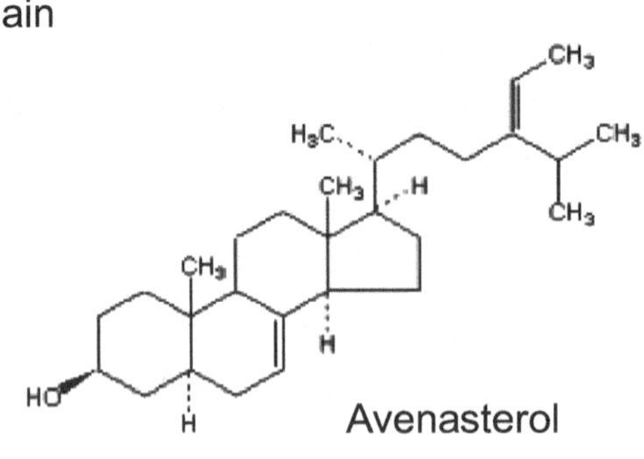

Avenasterol

Steroid biosynthesis

Steroids are made by the assembly and modification of isoprene units donated from the reactive precursors isopentenyl pyrophosphate and dimethylallyl pyrophosphate. These precursors can be made in different ways. In plants the non-mevalonate pathway uses pyruvate and glyceraldehyde 3-phosphate as substrates. One important reaction that uses these activated isoprene donors is steroid biosynthesis. Here, the isoprene units are joined together to make squalene and then folded up and formed into a set of rings to make lanosterol. Lanosterol can then be converted into other steroids such as cholesterol and ergosterol.

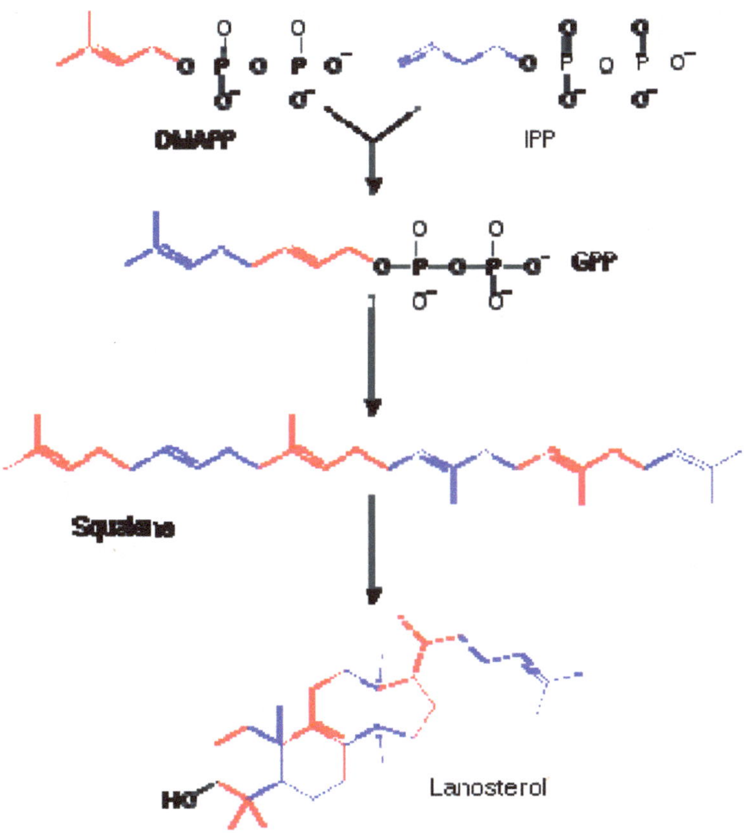

Fig. 2. Steroid synthesis pathway with the intermediates isopentenyl pyrophosphate (IPP), dimethylallyl pyrophosphate (DMAPP), geranyl pyrophosphate (GPP) and squalene

Chapter – 18

Alkaloids

Alkaloids are naturally occurring substances of plant origin that are alkaline in reaction. They are optically active, contain nitrogen as part of a heterocyclic ring and have varying degrees of psychological responses in man and animals. They exhibit significant pharmacological activity. Alkaloids are mostly basic in character but also includes some related compounds with neutral and even weakly acidic properties. Presently more than 2000 alkaloids are known.

Nomenclature of alkaloids

Alkaloids are named by adding the suffix 'ine' to various distinguishable features such as sources, physiological action in animals, their properties etc. Thus, 1. The name of the genera of the plants containing them Ex. Atropine (*Atropa*), nicotine (*Nicotiana*) *etc*. 2. The names of the species of the plants producing them Ex. Cocaine (*Erythroxolon coca*). 3. The common name of the organism producing them Ex. Ergotamine (ergot fungus). 4. The specific physiological activity observed in animals Ex. Emetine (emetic – causing vomiting), narcotine (narcotic-alters mind).

Classification of alkaloids

Various criteria are used for classifying alkaloids. Those alkaloids for which the precursors are the amino acids are grouped as True Alkaloids, while others are grouped as Pseudo alkaloids. Further, those alkaloids which contain nitrogen atom present in a side chain are called Proto alkaloids. However, the most widely used classification system of alkaloids is based on their skeleton rings.

1. "True alkaloids", contain nitrogen in the heterocyclic ring and originate from amino acids. Their characteristic examples are atropine, nicotine, and morphine. This group also includes some alkaloids that contain terpene (Ex. evonine) or peptide fragments (Ex.. ergotamine) in addition to the heterocyclic nitrogen ring. This group also includes piperidine alkaloids, coniine and coneine, although they do not originate from amino acids.
2. "Protoalkaloids", which contain nitrogen and also originate from amino acids. Examples include mescaline, adrenaline and ephedrine.
3. Polyamine alkaloids – derivatives of putrescine, spermidine, and spermine.
4. Peptide and cyclopeptide alkaloids.
5. Pseudalkaloids – alkaloid-like compounds that do not originate from amino acids. This group includes, terpene-like and steroid-like alkaloids, as well as purine-like alkaloids such as caffeine, theobromine and theophylline. Some compounds such as ephedrine and cathinone are included under pseudoalkaloids. They originate from the amino acid phenylalanine, but acquire their nitrogen atom not from the amino acid but through transamination.

Table 1. Classification of alkaloids based on chemical structure

S.No	Class	Examples
1.	Pyridine, piperidine	Conine, nicotine
2.	Tropane	Atropine, cocaine
3.	Quinoline	Quinine (*cinchona officinalis*)
4.	Isoquinoline	Papaverine (*papaver somniferum)*
5.	Phenanthrene	Morphine (opium poppy)
6.	Indole	Codeine,strychnine, ergotamine
7.	Purine	Caffeine (tea & coffee)
8.	Pyrrolidine	Hygrine (*Aswagandha*)
9.	Imidazole	Pilocarpine
10.	Phenylalylamine	Ephedrine
11.	Pyrolizidine	Retronecine
12.	Quinolizidine	Lupinine
13.	Steroidal	Salanidine
14.	Terpenoid	Aconotine

Based on the precursor amino acids also, the alkaloids can be classified. The precursor amino acids involved are tryptophan, lysine, ornithine, tyrosine, histidine and aspargine. The alkaloids derived from them are listed in Table 2. The biosynthetic pathways vary depending on the precursor compounds in the plant system.

Table 2. Classification of alkaloids based on amino acid precursors

S.No.	Precursor amino acids	Alkaloids
1.	Tryptophan Indoles	Quinolines
2.	Lysine	Piperidines
3.	Ornithine	Tropanes, Pyrrolidines
4.	Tyrosine	Isoquinolines
5.	Histidine	Phenylalkylamines
6.	Asparagine	Imidazoles, Pyridines

Functions of alkaloids in plants

The role of alkaloids in living organisms that produce them is not very clear. Alkaloids have been considered to be the final products of nitrogen metabolism in plants. Due to wide variations in the concentrations of alkaloids in plants, this thinking has been discarded. Alkaloids, however, have various functions in plants ranging from protection from animals to being sources of nitrogen and detoxification etc. They are listed below:

1. Act as protective agents against herbivorous animals due to their bitter taste and toxicity. The earlier adaptations like thorns and stinging hairs proved inadequate and then over the years, the plants learnt to produce phytochemicals like alkaloids for protection against animals. Possessing alkaloids thus became an evolutionary neccessity.

2. Some alkaloids act as the sources of nitrogen in case of deficiency of nitrogen. It is suggested that plants which cannot store excess nitrogen and can not afford to throw the element, temporarily store the excess nitrogen in the form of alkaloids, especially in vacuoles. It is redrawn whenever necessary.

3. Some alkaloids act as plant growth regulators.

4. Some alkaloids act as the final products of detoxification of various nitrogen containing substances.

Uses of alkaloids

Alkaloid-containing plants have been used by humans since ancient times for therapeutic and recreational purposes. For example, medicinal plants have been known in the Mesopotamia at least around 2000 BC. Extracts from plants containing toxic alkaloids, such as aconitine and tubocurarine, were used since antiquity for poisoning arrows.

1. **In medicine:** Medical use of alkaloid-containing plants has a long history. When the first alkaloids were isolated in the 19th century, they immediately found application in clinical practice. Many alkaloids are still used in medicine, usually in the form of salts. Alkaloids are particularly known for their therapeutic value. Accordingly, alkaloids find use in treating various health disorders. Some such uses are furnished in Table.3.

Table 3. Medicinal uses of alkaloids

S. No.	Alkaloid	Use
1.	Aconitine	Antipyretic
2.	Cocaine	Vasoconstrictor, local anesthesia
3.	Codeine	Cough suppressants
4.	Reserpine	Anti high blood pressures
5.	Emetine	Antiamoebic, emetic
6.	Morphine	Analgesic, narcotic
7.	Quinine	Antimalarial, antipyretic
8.	Ajmaline, quinidine	Antiarrhythmic
9.	Atropine	Boost nerve systems, anticholinergic
10.	Ergot alkaloids	Vasodilator, antihypertensive
11.	Vinblastine, vincristine	Antitumor
12.	Cholchicine	Gout remedy

Many synthetic and semisynthetic drugs are structural modifications of the alkaloids, which have been designed to enhance or change the primary effect of the drug and reduce unwanted side-effects. For example, naloxone, an opioid receptor antagonist, is a derivative of thebaine that is present in opium.

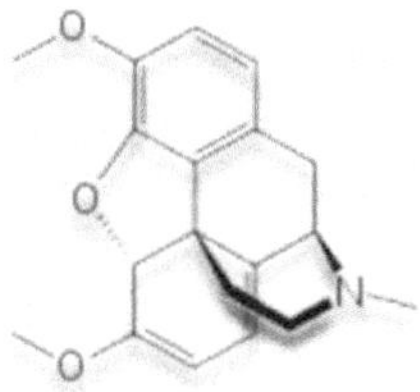

Thebaine

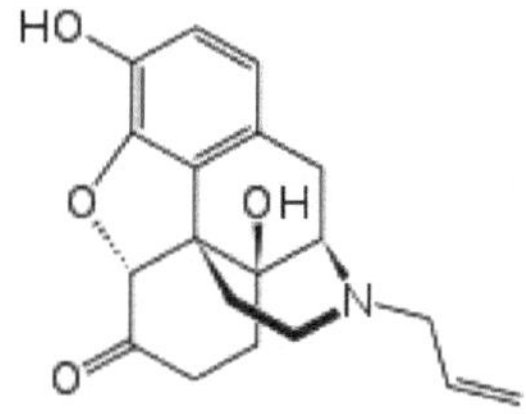

Naloxone

2. **As flavor and taste enhancing agents:** Capsaicin is the active ingredient of red chillies, and is used as flavor and taste enhancing agent.
3. **In Plant Biotechnology:** Colchicine is used to produce polyploidy in plants along with swollen and sterile seeds. Polyploid fruits and vegetables are now -a-days available commercially.
4. **As psychoactive agents:** Preparations of plants containing alkaloids and their extracts, and later pure alkaloids, have long been used in various ways. a. Stimulants: Ex. nicotine and caffeine, b. Hallucinogens: Ex. ergot alkaloids (LSD, lysergic acid and heroin), c. Tranquilizers: Ex. morphine and codeine. These alkaloids are popular pain killers:
5. **As pain killers:** Ex. aconitine and morphine
6. **As poisons:** Ex. coniine and stachydrine (*Medicago sativa*).
7. **As pesticides:** Some alkaloids, such as salts of nicotine and anabasine, have been used. Their use is getting reduced due to their high toxicity to humans.

Properties of alkaloids

Most alkaloids contain oxygen in their molecular structure. Such compounds are usually colorless crystals at ambient conditions. Oxygen-free alkaloids, such as nicotine or coniine, are typically volatile, colorless, oily liquids. Some alkaloids are colored, like berberine (yellow) and sanguinarine (orange).

Most alkaloids are weak bases, but some are amphoteric, for example theobromine and theophylline. Most alkaloids are poorly soluble in water but readily dissolve in organic solvents, such as diethyl ether, chloroform and 1,2-dichloroethane. However, caffeine dissolves well in boiling water. With acids, alkaloids form salts of various strengths. Those salts are usually soluble in water and alcohol and poorly soluble in most organic solvents. Exceptions include scopolamine hydrobromide, which is soluble in organic solvents and water-soluble quinine sulfate. Most alkaloids have a bitter taste. It is believed that plants evolved the ability to produce these bitter substances, many of which are poisonous, in order to protect themselves from animals. However, animals in turn have developed the ability to detoxify alkaloids.

Distribution in nature

The alkaloid content in plants is usually within a few percent and is inhomogeneous over the plant tissues. Depending on the type of plants, the maximum concentration is observed in the leaves (black henbane), fruits or seeds (Strychnine tree), root (*Rauwolfia serpentina*) or bark (*cinchona*). Furthermore, different tissues of the same plants may contain different alkaloids.

Extraction: Because of the structural diversity of alkaloids, there is no single method of their extraction from natural raw materials. Most methods exploit the property of alkaloid solublility in organic solvents. Their salts however, are water soluble.

Most plants contain several alkaloids. Their mixture is extracted first and then individual alkaloids are separated. Plant samples are thoroughly ground before extraction. Most alkaloids are present in the raw plants in the form of salts of organic acids. The extracted alkaloids may remain salts or change into bases. Base extraction is achieved by processing the raw material with alkaline solutions and extracting the alkaloid bases with organic solvents, such as 1,2-dichloroethane, chloroform, diethyl ether or benzene. Then, the impurities are dissolved by weak acids; this converts alkaloid bases into salts that are washed away with water. If necessary, an aqueous solution of alkaloid salts is again made alkaline and treated with an organic solvent. The process is repeated until the desired purity is achieved.

In the acidic extraction, the raw plant material is processed by a weak acidic solution (e.g., acetic acid in water, ethanol, or methanol). A base is then added to convert alkaloids to basic forms that are extracted with organic solvent (if the extraction is performed with alcohol, it is removed first, and the remainder is dissolved in water). The solution is later purified. Alkaloids are separated from their mixture using their different solubility in certain solvents and different reactivity with certain reagents or by distillation. Thin layer chromatography is an important technique used for separation of individual alkaloids.

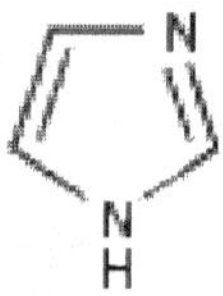

Imidazole

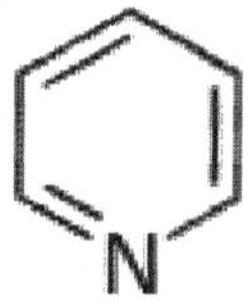

Pyridine

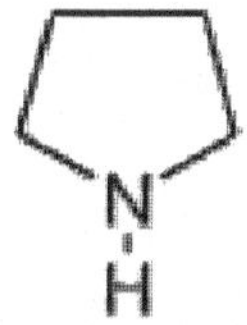

Pyrrolidine

Nitrogen containing heterocyclic compounds present in alkaloids

Some alkaloids

A widely known alkaloid is caffeine, a mild stimulant present in coffee beans, tea leaves and cola nuts. At around 200mg it stimulates the body. It is also useful as a stay awake item. It is mildly addictive with very limited negative effects. Cocaine is an alkaloid isolated from the cocoa leaves. It is a powerful stimulant. It increases stamina and reduces fatigue. Overdose may induce psychosis and hallucination. It was once used as an anesthetic.

The nicotine molecule contains both pyridine and pyrrolidine rings. It was isolated in 1828 by Posselt and Reimann. It is a liqiuid at room temperature with a boiling point of 246^0 C. It is levorotatory. It is miscible in water, alcohol and ether. The molecular formula is $C_{10}H_{14}N_2$. Nicotine on oxidation yields nicotinic acid, a vitamin.

Opium poppy cultivated all over the world is a source of many alkaloids. Opium contains 24 alkaloids, out of which 5 are important. They are morphine, codeine, narcotine, thebaine and papaverine. The first three are narcotic and analgesic. They have addictive properties. Morphine relieves from even severe pain. Codeine is used as a cough syrup.

Atropine in dilute solutions (as little as 4µg) causes dilatation of the pupil of the eye. Hence it is used prior to ophthalmic examination and eye surgery. Cocaine, similar in structure to atropine exhibits anesthetic properties.

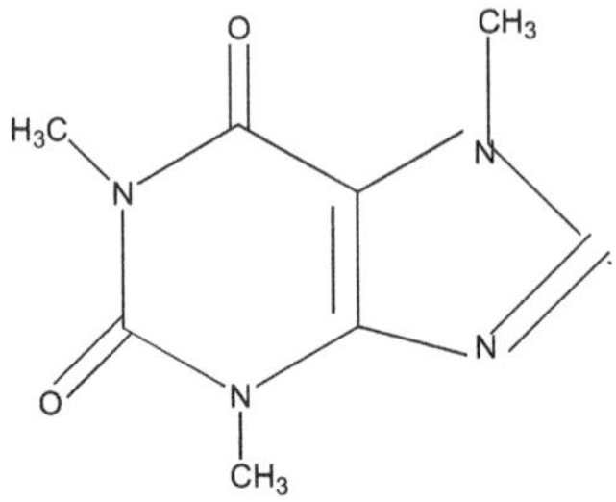

Caffeine

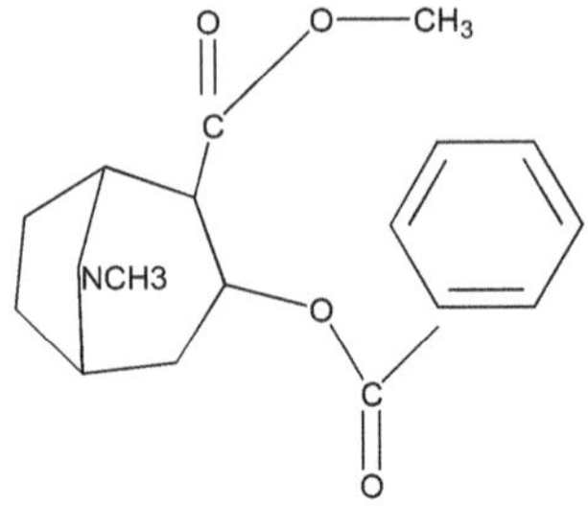

Cocaine

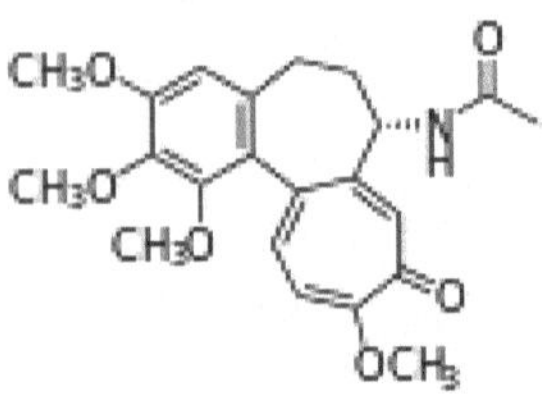

Colchicine

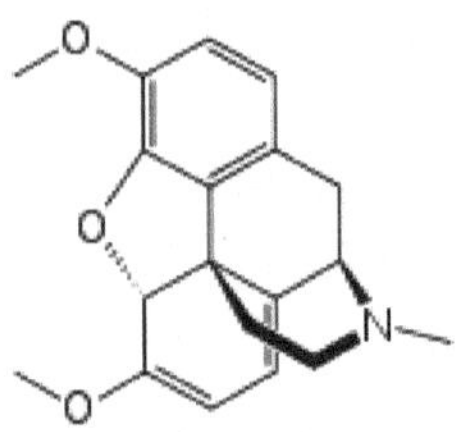

Thebaine

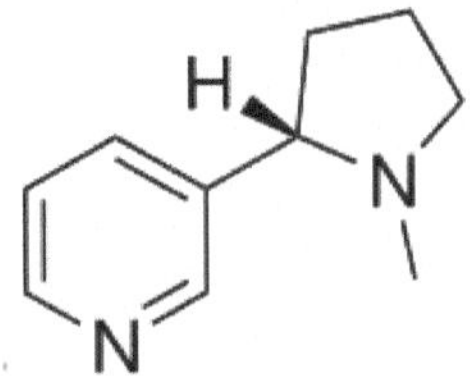

Nicotine

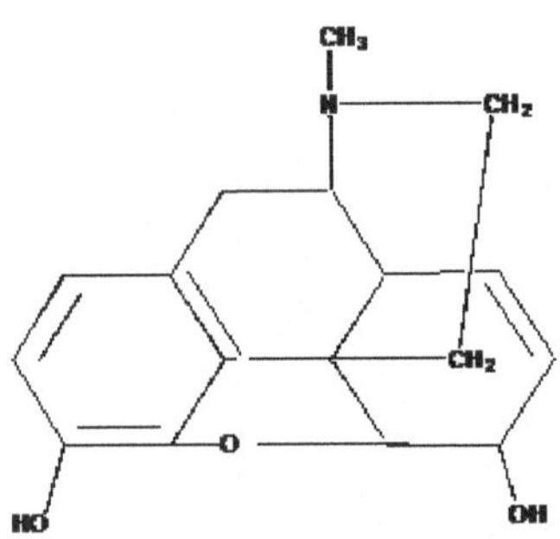

Morphine

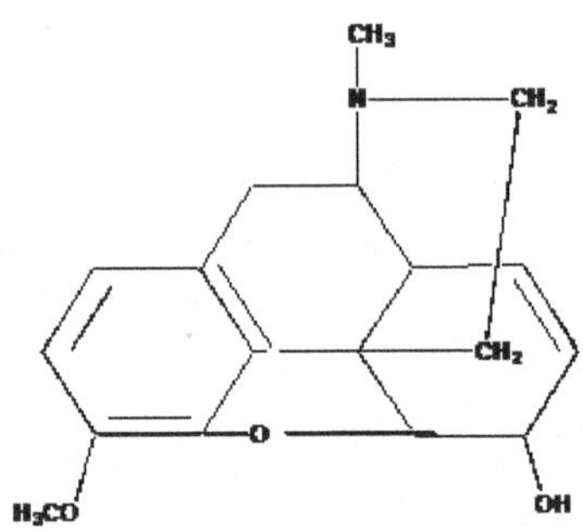

Codeine

Chapter – 19

Vitamins

Vitamins are essential organic nutrients required by an organism in small quantities. They are essential since they cannot be synthesized in sufficient quantities by an organism, and must be obtained from the diet. The term *vitamin* does not include other essential nutrients like dietary minerals, essential fatty acids, or essential amino acids which are needed in larger quantities than vitamins. The term *vitamin* is derived from "vitamine," coined in 1912 by the Polish biochemist Funk. The name is from *vital* and *amine,* meaning amine of life. Thirteen vitamins are generally recognized presently. Vitamins are used in small quantities by an organism to keep it in a healthy condition.

Vitamins are given letters of alphabet (A,B, C, D, etc.) in the order in which they have been discovered. The letters do not bear any relation to their structures or their importance. Vitamins are classified by their biological and chemical activity and not by their structure. Thus, each "vitamin" refers to a number of *vitamer* compounds that all show the biological activity associated with a particular vitamin. Vitamin A includes the compounds retinal, retinol, and four known carotenoids. Vitamers by definition are convertible to the active form of the vitamin in the body, and are sometimes inter-convertible to one another, as well.

Vitamins have diverse biochemical functions. Some have hormone-like functions as regulators of mineral metabolism (Ex. vitamin D), or regulators of cell and tissue growth and differentiation (Ex. some forms of vitamin A), others function like antioxidants (Ex. vitamin E and sometimes vitamin C). The largest number of vitamins (Ex. B complex vitamins) function as precursors for enzyme cofactors, that help enzymes in their work as catalysts in metabolism. As co-factors, vitamins may be tightly bound to enzymes as part of prosthetic groups. For example, biotin is part of enzymes involved in making fatty acids. Vitamins may also be less tightly bound to enzyme catalysts as coenzymes, detachable molecules that function to carry chemical groups or electrons between molecules. For example, folic acid carries various forms of carbon groups – methyl, formyl, and methylene – in the cell. These roles in assisting enzyme-substrate reactions are vitamins' best-known and most important functions.

Prior to extraction from yeast (vitamin B complex and semi-synthetic vitamin C) in mid 1930's vitamins were obtained solely through food intake. Such a dependence on diet greatly alters the type and amounts of vitamins ingested due to seasonal variations in diet and composition of vitamins etc. However, scope for resolution of the problem became possible after vitamins have been produced as commodity chemicals and made widely available as inexpensive semi-synthetic and synthetic dietary supplements and additives, since the middle of the 20th century. The information on the discovery and the sources from which they were recovered has been listed in Table 1.

Table 1. Disvovery and recovery sources of different vitamins

Year of discovery	Vitamin	Food source
1913	Vitamin A (Retinol)	Cod liver oil
1910	Vitamin B_1 (Thiamine)	Rice bran
1920	Vitamin C (Ascorbic acid)	Citrus, most fresh foods
1920	Vitamin D (Calciferol)	Cod liver oil
1920	Vitamin B_2 (Riboflavin)	Meat, eggs
1922	Vitamin E (Tocopherol)	Wheat germ oil, unrefined vegetable oils
1926	Vitamin B_{12} (Cobalamins)	liver, eggs, animal products
1929	Vitamin K_1 (Phylloquinone)	Leafy green vegetables
1931	Vitamin B_5 (Pantothenic acid)	Meat, whole grains, in many foods
1931	Vitamin B_7 (Biotin)	Meat, dairy products, eggs
1934	Vitamin B_6 (Pyridoxine)	Meat, dairy products
1936	Vitamin B_3 (Niacin)	Meat, eggs, grains
1941	Vitamin B_9 (Folic acid)	Leafy green vegetables

Classification of vitamins

Vitamins are classified into two groups based on their solubility as water-soluble or fat-soluble. Presently, there are 13 vitamins. Out of them four are fat-soluble (A, D, E, and K) and the remaining nine are water-soluble (8 B vitamins and vitamin C). Water-soluble vitamins dissolve easily in water and, in general, are readily excreted from the body. Since they are not readily stored in the body, they need to be regularly taken to replenish their requirement. They are polar compounds having –OH and –COOH groups. Good intake of fresh fruits and vegetables will satisfy the needs of water soluble vitamins. They are easily destroyed by heat, oxygen and U.V. light. Hence care should be taken while processing and storage of foods. These vitamins serve as cofactors to enzymes.

Fat-soluble vitamins are absorbed in the intestinal tract with the help of lipids (fats). They get accumulated in the body when taken in excess. These are non polar compounds and get deposited in fat deposits and membranes in the body. Hypervitaminosis results in the case of fat soluble vitamins and is one of the causes for cystic fibrosis. The different vitamins, their RDA(recommended daily allowance), deficiency disease and availability are presented in Tables. 2 and 3.

Table 2. Water soluble vitamins, needs, deficiency diseases and occurence

Common name	Chemical name	RDA	Deficiency disease	Occurrence/food source
Vitamin B_1	Thiamine	1.2mg	Beriberi	Oatmeal, brown rice, vegetables, liver, eggs
Vitamin B_2	Riboflavin	1.3mg	Antiriboflavinosis	Dairy products, bananas, popcorn, green beans, asparagus
Vitamin B_3	Niacin, niacinamide	16mg	Pellagra	Meat, fish, eggs, vegetable mushrooms, tree nuts
Vitamin B_5	Pantothenic acid	5mg	Paresthesia	Meat, broccoli, avocado
Vitamin B_6	Pyridoxine, pyridoxamine, pyridoxal	1.5mg	Anemia, peripheral neuropathy	Meat, vegetables, tree nuts, bananas
Vitamin B_7	Biotin	30μg	Dermatitis, enteritis	Raw egg yolk, liver, groundnuts, some vegetables
Vitamin B_9	Folic acid	400μg	Megaloblast, birth defects	Leafy vegetables, liver, cereals
Vitamin B_{12}	Cyanocobalamin	2.4μg	Megaloblastic anemia	Meat & animal products
Vitamin C	Ascorbic acid	90mg	Scurvy	Fruits, vegetables, liver

Table 3. Fat soluble vitamins, needs, deficiency diseases and occurence

Common name	Chemical name	RDA	Deficiency disease	Occurrence/ food source
Vitamin A	Retinol, retinal, & four carotenoids	900 μg	Night blindness, hyper keratosis	Orange, yellow fruits, pumpkin, spinach, carrots, leafy vegetables, liver
Vitamin D	Cholecalciferol	10 μg	Rickets, and osteomalacia	Fish, eggs, liver, mushrooms
Vitamin E	Tocopherols, tocotrienols	15mg	Hemolytic anemia (very rare)	Vegetable oils, fruits, vegetables
Vitamin K	Phylloquinone, menaquinone	120 μg	Bleeding diathesis	Leafy vegetables, eggyolk, liver,

In nutrition and disease

Vitamins are essential for the normal growth and development of most of the organisms. Vitamins and minerals are required specifically depending on the stage of development. These nutrients facilitate the chemical reactions that produce different parts of the body like skin, bone, and muscle. Deficiency develops if there is insufficient supply of one or the other vitamin or mineral. Even minor deficiencies may cause permanent damage.

Vitamins are mostly obtained through food, but a few are obtained by other means. For example, intestinal microorganisms (gut flora) produce vitamin K and biotin, while vitamin D is synthesized in the skin with the help of the ultraviolet rays from the sunlight. Humans can produce some vitamins from precursors they consume. Examples include vitamin A, produced from beta carotene, and niacin, from the amino acid tryptophan.

Once growth and development are completed, vitamins remain essential nutrients for the healthy maintenance of the cells, tissues, and organs that make up a multicellular organism; they also enable a multicellular life form to efficiently use chemical energy provided by the food it consumes, and to help process the proteins, carbohydrates, and fats required for respiration.

Deficiencies

Human beings must consume vitamins to avoid deficiency. The human body's stores for different vitamins vary widely. Vitamins A, D, and B_{12} are stored in significant amounts in the human body, mainly in the liver, and an adult human's diet may be deficient in vitamins A and D for many months and B_{12} in some cases for years, before developing a deficiency condition. However, vitamin B_3 (niacin and niacinamide) is not stored in the human body in significant amounts, so stores may last only a couple of weeks. It may take a long time for the deficiency symptom to express itself. For example, vitamin C deficiency and the first symptoms of scurvy take from a month to more than six months to express.

Deficiencies of vitamins are classified as either primary or secondary. A primary deficiency occurs when an organism does not get enough of the vitamin in its food. A secondary deficiency may be due to an underlying disorder that prevents or limits the absorption or use of the vitamin, due to a "lifestyle factor", such as smoking, excessive alcohol consumption, or the use of medications that interfere with the absorption or use of the vitamin. People who eat a varied diet are unlikely to develop a severe primary vitamin deficiency. In contrast, restrictive diets have the potential to cause prolonged vitamin deficits, which may result in often painful and potentially deadly diseases.

Well-known human vitamin deficiencies involve thiamine (beriberi), niacin (pellagra), vitamin C (scurvy), and vitamin D (rickets). In the developed world, such deficiencies are rare. This is due to an adequate supply of food and the addition of vitamins and minerals to common foods, often called fortification

Dietary supplements, often containing vitamins, are used to ensure that adequate amounts of nutrients are obtained on a daily basis. Scientific evidence supporting the benefits of some vitamin supplements is well established for certain health conditions. In some cases, vitamin supplements may have unwanted effects, especially if taken before surgery, with other dietary supplements or medicines, or if the person taking them has certain health conditions. Dietary supplements may also contain levels of vitamins many times higher, and in different forms, than one may ingest through food.

Vitamin A

Vitamin A is found in milk, eggs, oils and vegetables. Carrots are rich in vitamin A. It stimulates growth and is useful for good vision. Vitamin A is part of rhodopsin, the pigment present in the eye. Deficiency of this vitamin causes night blindness, reduce immune response and inhibits growth. Another name for this vitamin is retinal. It is fat soluble. Larger doses of this vitamin are not advised. The upper limit of the recommended daily allowance of vitamin A is 3mg. Yellow and green fruits like carrots, apricots, broccoli, papaya, peaches and spinach are rich in vitamin A. β carotene from plants is converted to vitamin A in the liver.

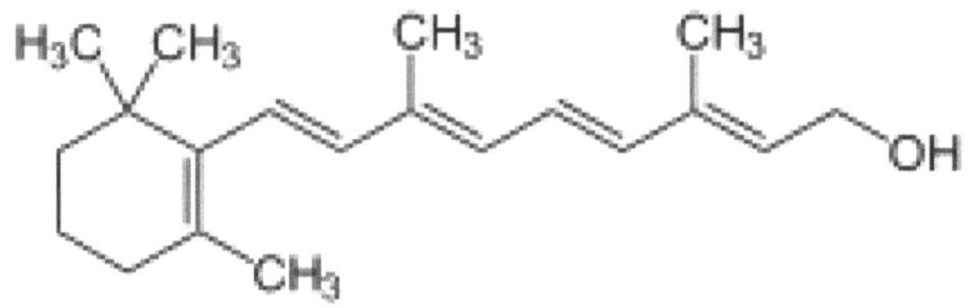

Vitamin A

Vitamin B group

This B group contains many vitamins and are generally water soluble. They are very important and need to be regularly supplied to the body through a balanced diet. The vitamins in this group are B_1, B_2, B_3, B_6, B_7, B_9 and B_{12}. These vitamins do not have any common structure. All of them function as cofactors to different enzymes and help release energy from foods. They are useful in maintaining healthy skin, muscles and good health.

Vitamin B_1, thiamine, helps carbohydrate to release energy. Deficiency of this vitamin leads to a disease called 'beriberi'. It causes muscle and heart weakness. Deficiency symptoms are fatigue, poor appetite and weight loss. Vitamin B_1 is present in liver, kidney, yeast, eggs, green vegetables, cereals and fresh fruit. The RDA for adults is about 1.2mg.

Riboflavin (Vitamin B_2), is a co-enzyme useful in break down of fats, sugars and proteins. It also has a role in synthesis of mucous material. Deficiency of this vitamin leads to cracking of lips, dry skin, dermatitis, tongue inflammation and cataract. It is present in animal food, whole grains and dark green leafy vegetables. Riboflavin is needed for good vision, healthy skin and good hair. The RDA for aduls is 1.3 mg.

Vitamin B_3 (niacin) also has a role in release of energy from food. Insufficiency causes a disease called 'pellagra', which looks like blisters on the skin. In extreme conditions it affects the nervous system, causing mental disturbances. It has a useful role in reducing cholesterol in the blood. The name niacin has been derived from its actual name nicotinic acid. The RDA for adults is 13-18mg. Niacin can be obtained from chicken, beef, fish, liver, brown rice and whole grains.

Pantothenic acid or vitamin B_5 is part of coenzyme A, which transfers acetyl group from pyruvate to citric acid for the production of energy. It is also involved in the breakdown of fats and fatty acids. Intestinal bacteria provide this vitamin to the body. The RDA is 1-2 mg. Deficiency causes fatigue, retarded growth, muscle cramps and anemia. It is available in salmon fish, beef, eggs, brewer's yeast, whole grains and fresh vegetables.

Pyridoxine or vitamin B_6 is required in the uptake and metabolism of amino acids, fats and formation of red blood cells. A balanced meat, fruit and cereal diet makes available sufficient levels of this vitamin. The RDA is about 2mg. Deficiency leads to dermatitis, fatigue, anemia, kidney stones and retarded growth.

Vitamin B_{12} or cobalamin or cyanocobaltamin has the most complicated chemical structure. It has one cobalt atom in its molecule with four pyrrole rings around it. Cobalamine participates in the transfer of methyl groups, molecular rearrangements, formation of red blood cells and synthesis of acetyl choline. The RDA is 3 μg for adults. Deficiency results in pernicious anemia. Dietary sources are liver, kidney, chicken, fish, curds and milk. Cobalamin is not present in plants and hence vegetarians need this vitamin as a supplement.

Folic acid or vitamin B_9 is needed for the formation of hemoglobin. Fresh foods contain this vitamin while stored foods do not contain this vitamin. Since it is water soluble it needs to be replenished regularly. The RDA is 0.4mg for adults. Deficiency leads to abnormal red blood cells, anemia, loss of hair and depression. It is present in leafy (folium, hence folic acid) vegetables, beans, sea food, whole grains and asparagus.

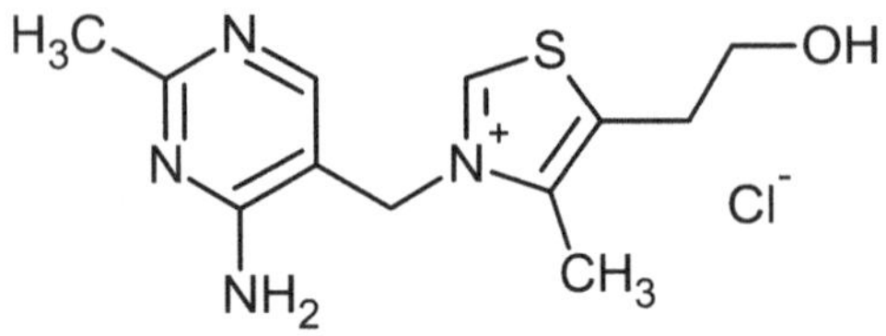

Vitamin B1 Thiamine

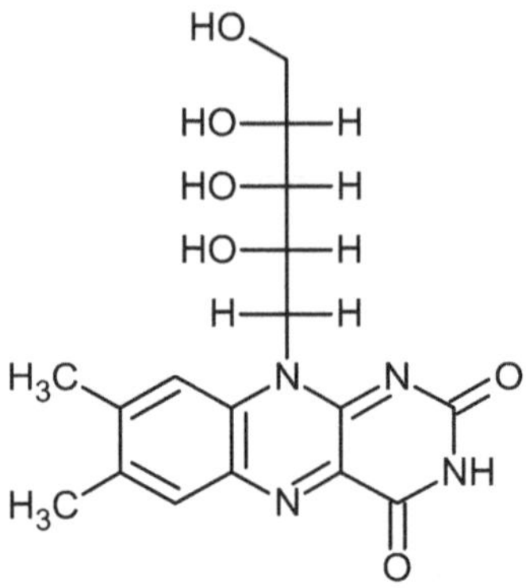

Vitamin B2 Riboflavin

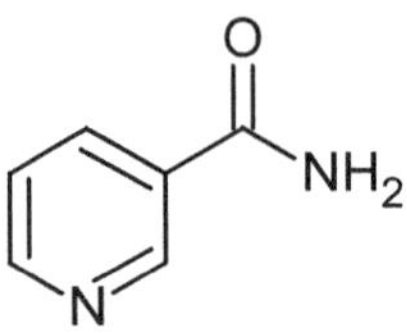

Vitamin B3 Niacin

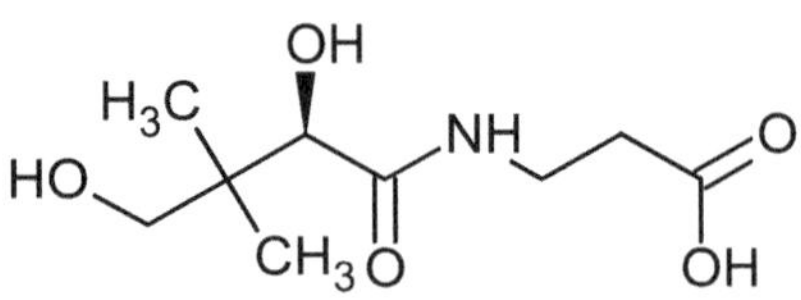

Vitamin B5 Pantothenic acid

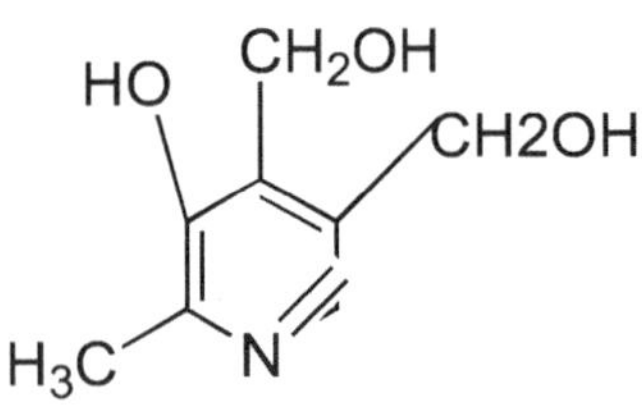

Vitamin B6 Pyridoxine

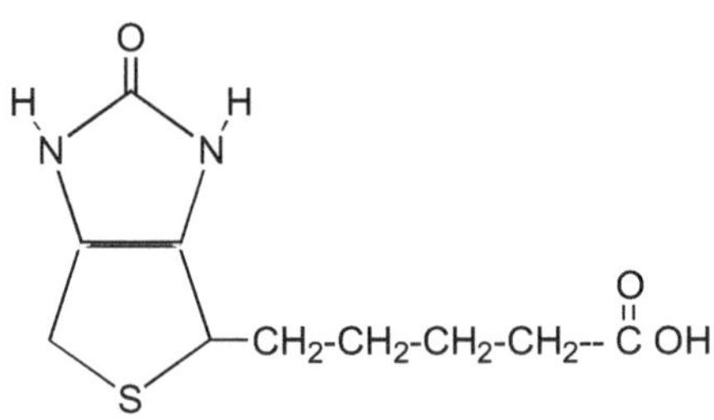

Vitamin B7 Biotin

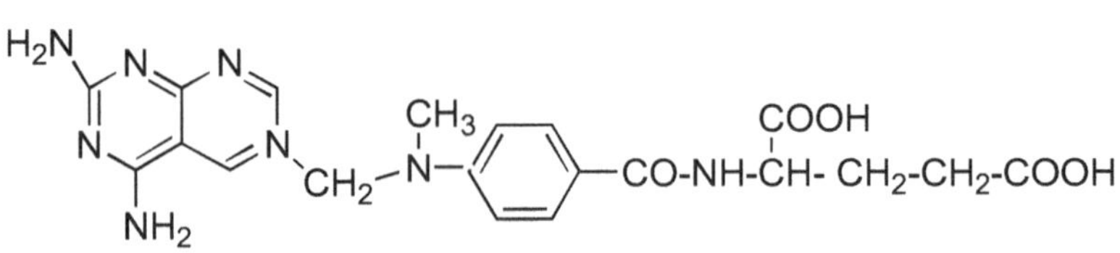

Vitamin B9 Folic acid

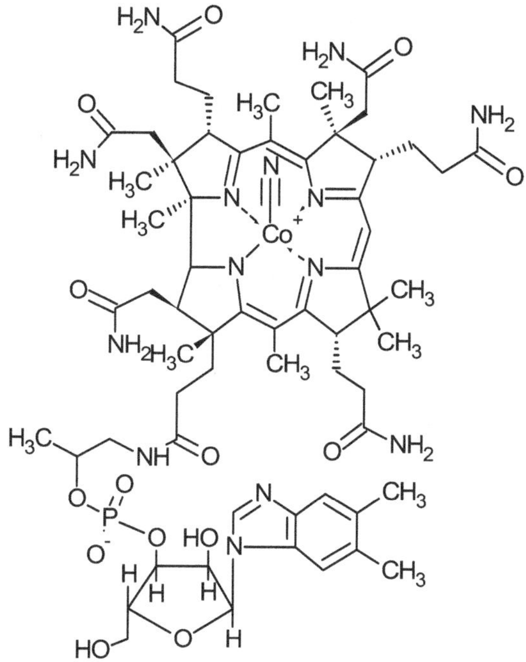

Vitamin B12 Cobalamin

Vitamin C

Vitamin C or ascorbic acid is a water soluble molecule. Deficiency of this vitamin results in a disease called 'scurvy' leading to bleeding gums, weak connective tissue, slow healing of wounds and anemia. A regular supply of this vitamin is essential for skin health and in general, good health. Vitamin C occurs in almost all fruits, vegetables and sprouts. Since it is water soluble it needs to be replenished to the body. Its major function is in the synthesis of hydroxyproline and hydroxy lysine, which are needed to make collagen. This protein is part of the tendons, connective tissue, bone structure, and skin. The RDA of this vitamin is 90mg.

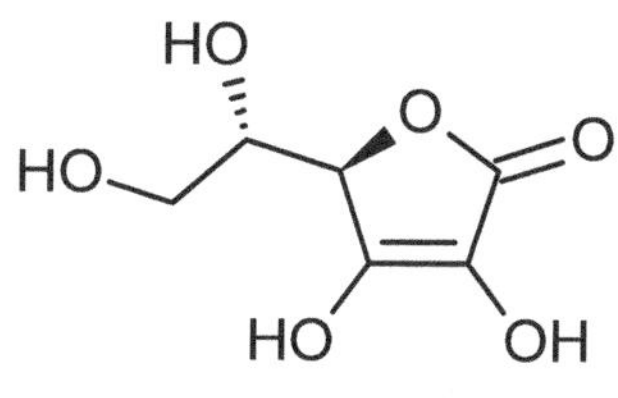

Vitamin C Ascorbic acid

Vitamin D series

Two vitamins, namely, vitamins D_2 and D_3 are fat soluble and can be stored in the body. Vitamin D_2, or calciferol can be obtained from foods like vegetables, egg yolk, cod liver oil and milk products while D_3 or cholecalciferol is synthesized in the body with the help of sunlight. D_3 is synthesized by the skin with the help of ultra violet light from the sun. Low vitamin D can lead to bone disorders like osteoporosis and rickets. About 10mg of vitamin D is required by the body. Exposure to sun even for a small duration is beneficial. Vitamin D helps in metabolizing calcium and phosphorus effectively. In cold temperate countries, where exposure to sun is very limited, vitamin D supplements are essential.

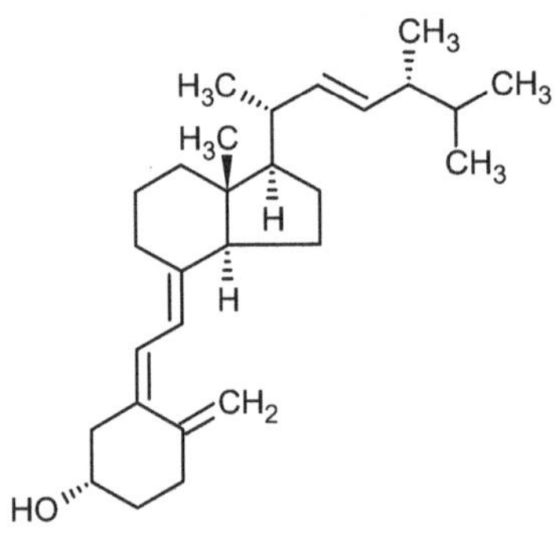

Vitanin D2 Calciferol

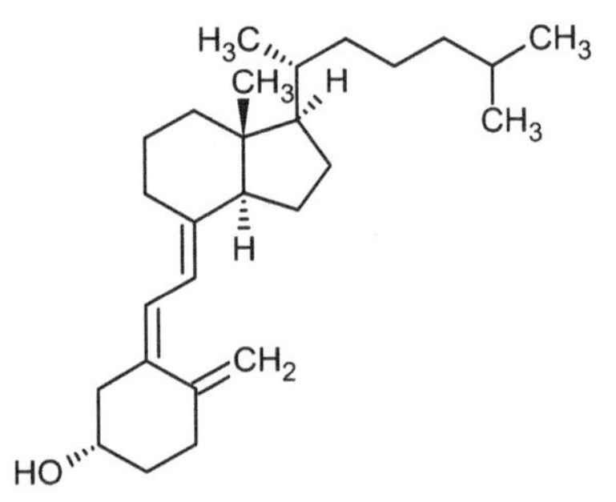

Vitamin D3 Cholecalciferol

Other vitamins

Vitamins E, H and K are the other important vitamins needed by our body, in smaller quantities. Balanced diet normally takes care of their requirement.

Vitamin E (tocopherols) act as antioxidants and prevent free radical formation. Vegetable oils serve as good sources of this vitamin. The RDA for adults is about 10 -15 mg. A deficiency of tocopherols causes hemolysis and anemia. Vitamin E is available in whole grains, vegetables and vegetable oils.

Vitamin H (or B 7) or biotin is a coenzyme involved in transfer of carboxyl groups during synthesis of fatty acids. The RDA is 30µg. The deficiency of biotin is very rare. However deficiency can cause dermatitis, loss of hair, anemia and depression. Dietary sources of biotin include liver,yeast,nuts and eggs.

Vitamin K_1 or phylloquinone is found in plants. Vitamin K_2 or menadione, is found in spinach and cabbage and also in animals. K_1 has a large saturated side chain while K_2 has a very large unsaturated side chain. It acts as a blood clotter. Zymogen needed for the blood clotting process is synthesized with the help of vitamin K_2. The RDA for vitamin K is 80 micrograms. Deficiency causes bleeding from minor cuts, delayed clotting times and bruises.

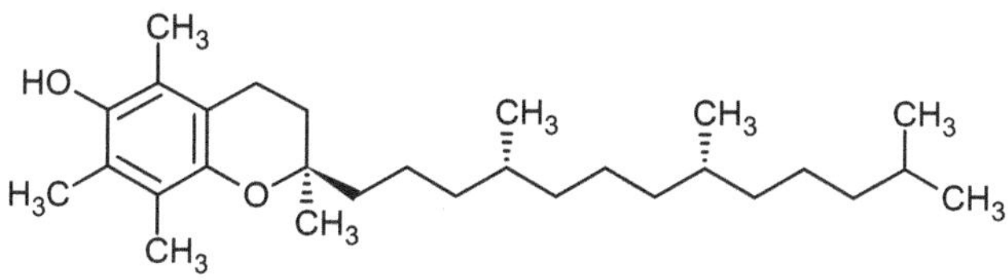

Vitamin E Tocopherol

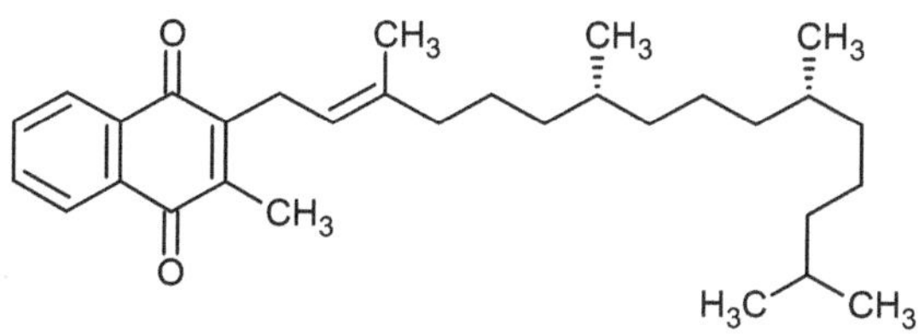

Vitamin K1 Phylloquinone

Chapter – 20

Plant Growth Regulators

Plant hormones (phytohormones) are chemicals that regulate plant growth and are termed as 'plant growth substances'. Plant hormones are signal molecules produced within the plant, and are present in extremely low concentrations. Hormones regulate cellular processes in targeted cells. Hormones also determine the formation of flowers, stems, leaves, the shedding of leaves, and the development and ripening of fruit. Every cell is capable of producing hormones in plants. They shape the plant, affecting seed growth, time of flowering, the sex of flowers, senescence of leaves, and fruits. Plant hormones affect gene expression and transcription levels, cellular division, and growth.They direct which tissues grow upward and which grow downwards. Hormones are vital to plant growth. Hence, they are also known as growth factors or growth regulators.

The word hormone is derived from Greek, meaning *set in motion*. Plant hormones are not nutrients, but chemicals that in small amounts promote and influence the growth, development, and differentiation of cells and tissues. The concentration of hormones required for plant responses are very low (10^{-6} to 10^{-5} mol/l).

Five major classes of plant hormones have been identified. They are abscisic acid, auxins, cytokinins, ethylene and gibberellins. These are more widely occurring and influence the plant growth in many ways. In addition, there are a few more minor plant hormones. The five major classes of phytohormones are first discussed followed by a brief mention of the minor ones.

Abscisic acid

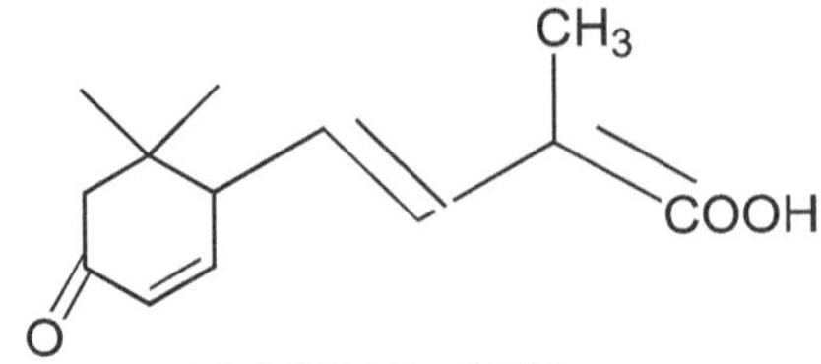

ABSCISIC ACID

Abscisic acid (ABA) is an important plant growth regulator The name "abscisic acid" was given because it was found in high concentrations in newly abscissed or freshly fallen leaves. It is produced in the leaves of plants, originating from chloroplasts, especially when plants are under stress. In general, it acts as an inhibitory chemical compound that affects bud growth, and seed and bud dormancy. Plants start life as a seed with high ABA levels. Just before the seed germinates, ABA levels decrease. During germination and early growth of the seedling, ABA levels decrease even more. As plants begin to produce shoots with fully functional leaves, ABA levels begin to increase, slowing down cellular growth in more "mature" areas of the plant. Stress from water or predation affects ABA production and catabolism rates, mediating another cascade of effects that trigger specific responses from targeted cells. Scientists are still piecing together the complex interactions and effects of this and other phytohormones.

Auxins

The main auxin is indole-3-acetic acid. Auxins are compounds that positively influence cell enlargement, bud formation and root initiation. They also promote the production of other hormones and in conjunction with cytokinins, they control the growth of stems, roots, and fruits, and convert stems into flowers. Auxins act to inhibit the growth of buds lower down the stems (apical dominance), and also to promote lateral and adventitious root development and growth. Leaf abscission is initiated by the growing point of a plant ceasing to produce auxins. Auxins in seeds regulate specific protein synthesis as they develop within the flower after pollination, causing the flower to develop a fruit to contain the developing seeds.

Auxins are toxic to plants in large concentrations. They are most toxic to dicots and less so to monocots. Because of this property, synthetic auxin herbicides like,2,4-D (2,4-dichlorophenoxy acetic acid) have been developed and used for weed control. Auxins, especially 1-naphthaleneacetic acid (NAA) and Indole-3-butyric acid (IBA), are also commonly applied to stimulate root growth when taking cuttings of plants. The most common auxin found in plants is indole-3-acetic acid or IAA.

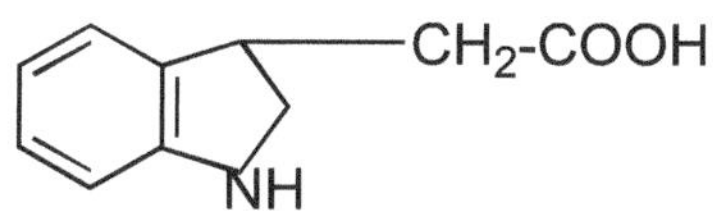

INDOLE 3- ACETIC ACID

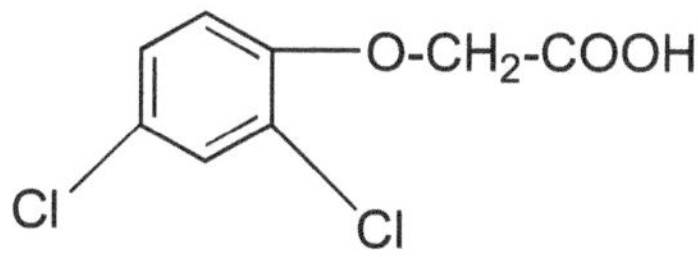

2,4, D a synthetic auxin

Cytokinins

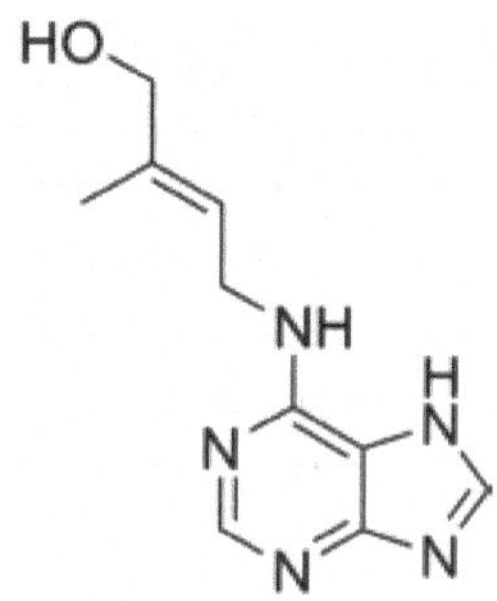

Zeatin

Cytokinins (CKs) are a group of chemicals that influence cell division and shoot formation. They also help delay senescence or the aging of tissues. They have a highly synergistic effect in concert with auxins. Cytokinins counter the apical dominance induced by auxins. Along with ethylene, CKs promote abscission of leaves, flower parts, and fruits. The correlation of auxins and cytokinins in the plants is a constant . Zeatin is the first CK that has been isolated from the immature kernels of *Zea*. Adenine or adenine like compounds induce cell division. All CKs are similar to the adenines. CK was discovered in 1964. CKs are produced in the roots,seeds, fruits and young leaves.

Ethylene

Ethylene is a gas that forms through the breakdown of methionine and is present in all cells. Ethylene is insoluble in water and does not accumulate within the cell but diffuses out of the cell and the plant. Ethylene affects cell growth and cell shape. When a growing shoot hits an obstacle underground, ethylene production greatly increases, preventing cell elongation and causing the stem to swell. The resulting thicker stem can exert more pressure against the object impeding its path to the surface. When stems of trees are subjected to wind, causing lateral stress, greater ethylene production occurs, resulting in thicker, more sturdy tree trunks and branches. Ethylene affects fruit-ripening. Normally, when the seeds are mature, ethylene production increases and builds-up within the fruit, resulting in a climacteric event just before seed dispersal. Ethylene inhibits flowering in most species of plants, but promotes in pine apple and mango plants.

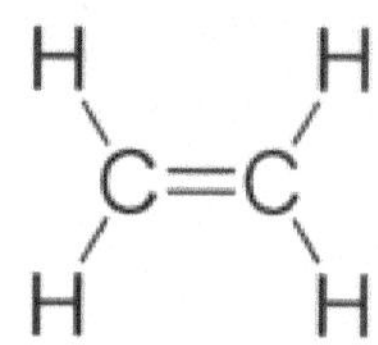

Ethylene

Gibberellins

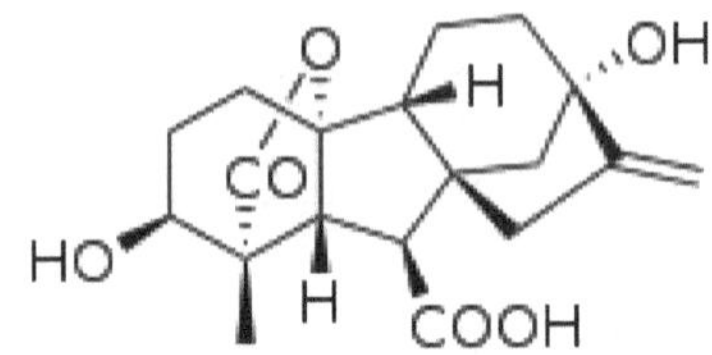

Gibberellin A1

Gibberellins, or GAs, include a large range of chemicals that are produced naturally within plants and by fungi. They were first discovered when a chemical produced by a fungus called *Gibberella fujikuroi*, produced abnormal growth in rice plants. Gibberellins are important in seed germination, affecting enzyme production that mobilizes food production used for growth of new cells. This is done by modulating chromosomal transcription. In grain (rice, wheat, corn, etc.) seeds, a layer of cells called the aleurone layer wraps around the endosperm tissue. Absorption of water by the seed causes production of GA. The GA is transported to the aleurone layer, which responds by producing enzymes that break down stored food reserves within the endosperm, which are utilized by the growing seedling. GAs produce bolting of rosette-forming plants, increasing internodal length. They promote flowering, cellular division, and in seeds growth after germination. Gibberellins also reverse the inhibition of shoot growth and dormancy induced by ABA. Gibberellins are synthesized in apical portions of stems and roots.

Other known hormones

The following are some more plant growth regulators. These also occur in small quantities and have important specific roles in plant growth regulation.

- Brassinosteroids, are a class of polyhydroxysteroids and have been recognized as a sixth class of plant hormones, which stimulate cell elongation and division, gravitropism, resistance to stress, and xylem differentiation. They inhibit root growth and leaf abscission. Brassinolide was the first identified brassinosteroid and was isolated from extracts of rapeseed (*Brassica napus*) pollen in 1979.
- *Salicylic acid:* activates genes in some plants that produce chemicals that aid in the defense against pathogenic invaders.
- *Jasmonates*: are produced from fatty acids and promote the production of defense proteins that are used to fend off invading organisms. They seem to have a role in seed germination, and affect the storage of protein in seeds and also affect root growth.

- *Plant peptide hormones*: are small secreted peptides that are involved in cell-to-cell signaling. These small peptide hormones play crucial roles in plant growth and development, including defense mechanisms, the control of cell division and expansion, and pollen self-incompatibility.
- *Polyamines:* are strongly basic molecules with low molecular weight that are present in most organisms. They are essential for plant growth and development and affect the process of mitosis and meiosis.
- *Nitric oxide (NO)*: serves as signal in hormonal and defense responses (Ex. stomatal closure, root development, germination, nitrogen fixation, cell death and stress response). NO might be produced by an undefined NO synthase, a special type of nitrite reductase, nitrate reductase, mitochondrial cytochrome c oxidase or non enzymatic processes and regulate plant cell organelle functions (Ex. ATP synthesis in chloroplasts and mitochondria).
- Strigolactones are implicated in the inhibition of shoot branching.
- Karrikins, a group of plant growth regulators are present in the smoke of burning plant material. They have the ability to stimulate the germination of seeds.

Table 1. Important plant growth regulators and their properties

Class	Function (s)	Practical uses
Auxins	Shoot elongation	Thin tree fruit, increase rooting and flower formation
Gibberellins	Stimulate cell division and elongation	Increase stalk length, increase flower and fruit size
Cytokinins	Stimulate cell division	Prolong storage life of flowers and vegetables and stimulate bud initiation and root growth
Ethylene generators	Ripening	Induce uniform ripening in fruit and vegetables
Growth inhibitors	Stops growth	Promote flower production by shortening internodes
Growth retardants	Slows growth	Retard tobacco sucker growth

Chapter – 21

Nitrogen Metabolism

Before the invention of N fertilizer, mineralization of organic N (present in dead plants and animals or manure) was the main method used to obtain ammonium and nitrate nitrogen. Mineralization is conducted by soil microbes resulting in ammonium ion. Most Nitrogen is absorbed by plants in the form of Nitrate (NO^-_3). Ammonium is converted to Nitrate with the help of soil bacteria. Nitrogen is a very mobile element and is easily lost from the soil. Ammonium (NH_4^+) can be converted to Ammonia (NH_3). The gaseous ammonia can be lost to the atmosphere. Nitrate can be lost through denitrification. Nitrate nitrogen is reduced to nitrous oxide (N_2O), elemental nitrogen (N_2) and nitric oxide (NO). All are volatile. Denitrification occurs under conditions of high water availability and temperatures higher than 10^0 C.

Ammonia is manufactured by the Haber process by combining nitrogen and hydrogen gases under high pressure and temperatures. This is the primary step in the manufacture of nitrogenous fertilizer, an essential nutrient required for crops. Nitrogen is also fixed biologically. This process does not require high temperature or pressures. The enzyme nitrogenase can enable this process at ambient temperatures more efficiently. The nitrogen fixation is two fold more efficient than the inorganic Haber process. An energy requirement of 355KJ/ mole of NH_3 is required.

$$N2 + 6H+ + 6e- \longrightarrow 2NH3 \quad \text{Reduction reaction}$$

Nitrogen fixation

Nitrogen fixation is the process by which atmospheric nitrogen gas is converted into ammonia. The basic cellular requirements for the biological reduction of nitrogen are 1).an active nitrogenase enzyme, 2). Mg^{++} and a continuous supply of ATP, 3). A strong reducing agent that is a suitable electron donor and 4). Low oxygen tension. The ammonia is subsequently available for many important biological molecules such as amino acids, proteins, vitamins, and nucleic acids. The reaction with ATP and other reducing systems can be presented as follows:

$$N2 + 16\ ATP + 8e- + 2H+ \longrightarrow NH3 + 16\ ADP + 16\ Pi + H2$$

Nitrogenase enzyme

Nitrogenase enzyme consists of two oxygen sensitive non-haem iron proteins. The proteins are brown in colour due to the presence of iron. Both of them are not active in nitrogen fixation separately. The larger protein (molybdo ferredoxin) consists of four sub-units of two different types and contains Mo in addition to Fe. The smaller protein (azoferredoxin) contains only Fe and contains two identical sub-units. These two proteins are present in 1:4 proportion. Ammonia is the only stable product of nitrogen fixation. Nitrogenase enzyme was discovered in 1960 by Cornaham and his collaborators while working with an extract of *clostridium pasteurianum*.

Table 1. Components of nitogenase system

Property	Azoferredoxin	Molybdoferredoxin
Molecular weight	55,000	220,000
No. of sub-units	2	2x2
Fe atoms	4	18
Mo atoms	0	8
Acid labile sulphide	11	16

ATP requirements

The ATP required for nitrogen fixation in physiological terms is a bioenergenetically expensive process. The ATP is supplied by 1). Oxidation of respiratory substrates, 2). Through conversion of pyruvate to acetate and 3). Photosynthetic systems via photophosphorylation.

Reductant requirement

Ferredoxins and flavodoxins are small specialized proteins. They serve as strong reducing agents. Ferredoxins are involved in many biological processes like photosynthesis and pyruvate metabolism. Flavodoxins are yellow proteins in an oxidized state.

NADPH → Ferredoxin → Azotoflavin

Fe-protein (Nitrogenase) → Mo-Fe protein (Nitrogenase) → N2

Restriction of oxygen

Nitrogen fixation normally takes place under anaerobic conditions or when oxygen is limiting. The symbiotic N fixation in root nodules takes place with the help of leghaemoglobin. This enzyme regulates oxygen concentration. Leghemoglobin, is an important oxygen-binding protein present in nodules. It delivers oxygen to the *rhizobia*, the nitrogen fixing bacteria, at a concentration, harmless to the nitrogenase enzyme. Root nodule is a highly specialized compartment in which nitrogen fixation and oxidative metabolism take place.

ATP- dependent hydrogen evolution

Experimental studies have revealed that for nitrogen fixation, 75% of the available electrons are utilized. The rest of the electrons are used to reduce proton (hydrogen ion) in an ATP- dependent process catalyzed by nitrogenase enzyme and Mg^{++}. Many nitrogen fixing bacteria are unable to prevent this hydrogen production step. This accounts for the low efficiency of nitrogen fixation. Non-hydrogen evolving symbiotic systems are more efficient in nitrogen fixation. Non- leguminous symbiotic N-fixation appears to be more efficient.

Functions of the two proteins

For nitrogenase to function, a complex between the iron protein and the Mo- Fe protein must form. The most reduced and predominant form exists when all the above conditions are satisfied. The Mo-Fe protein reacts with the substrates like nitrogen, acetylene and other reducible components. In essence the N-fixation involves the reduction of Fe atoms in the Fe- protein due to which ATP is hydrolysed to ADP and Phosphate. An electron from the Fe atoms in the Mo-Fe protein is used to reduce the substrate. Many such steps are needed before the final reduced product namely, NH_3 is formed .

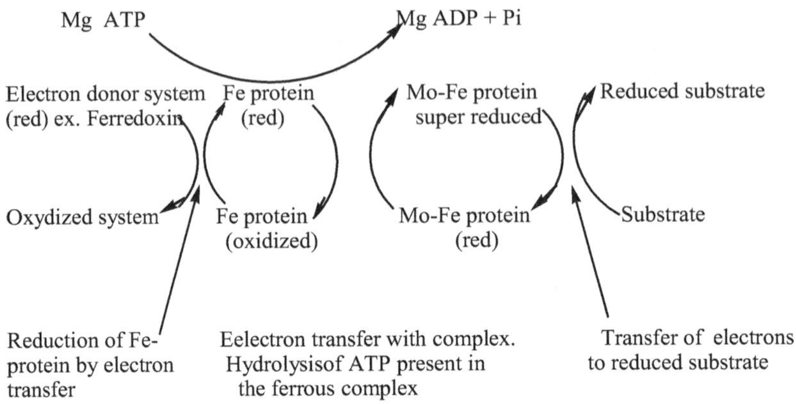

REDUCTION OF SUBSTRATES BY NITROGENASE

Organisms that can fix nitrogen

Cyanobacteria found in a variety of terrestrial and aquatic habitats are capable of nitrogen fixation.The aquatic fern Azolla is the only fern that can fix nitrogen. It does so by virtue of a symbiotic association with a cyanobacterium (*Anabaena azollae*). Azolla is found worldwide and is sometimes used as a valuable source of nitrogen for agriculture. Alder is a N-fixing plant that forms a symbiotic association with bacteria. Gunnera sp., is an unusual angiosperm that contains N-fixing cyanobacteria in pockets at the base of petioles. This is the only angiosperm that forms a N-fixing symbiosis with cyanobacteria. Nodules present on plants like groundnut, soybean and other bean plants fix nitrogen with the help of leghemoglobin, an important oxygen-binding protein.

Nitrogen cycle

Nitrogen is important in forming the amino portions of the amino acids which in turn form the proteins and all enzymes. The chief reservoir of nitrogen is the atmosphere, which is about 78% nitrogen. Nitrogen gas can be taken from the atmosphere (fixed) in two basic ways. First, lightning provides enough energy to "burn" the nitrogen and fix it in the form of nitrate. The other form of nitrogen fixation is by nitrogen fixing bacteria, who use special enzymes instead of the extreme amount of energy found in lightning to fix nitrogen. These nitrogen-fixing bacteria come in three forms: some are free-living in the soil; some form symbiotic, mutualistic associations with the roots of bean plants and other legumes (rhizobial bacteria); and the third form of nitrogen-fixing bacteria are the photosynthetic cyanobacteria (blue-green algae) which are found most commonly in water. All of these fix nitrogen, either in the form of nitrate or in the form of ammonia.

Most plants can take up nitrate and convert it to amino acids. When plants or animals die (or release waste) the nitrogen is returned to the soil. The usual form in which nitrogen is returned to the soil is through ammonia which is rather toxic. But, fortunately, there are nitrite bacteria in the soil and in the water which take up ammonia and convert it to nitrite. Nitrite is also somewhat toxic, but another type of bacteria, nitrate bacteria, take nitrite and convert it to nitrate, which can be taken up by plants to continue the cycle. There are denitrifying bacteria which take the nitrate and combine the nitrogen back into N_2 gas.

Nitrate and ammonium transport

The nitrate uptake system of higher plants consists of a constitutive, low affinity transport system (LATS) and a high affinity transport system (HATS) regulated by cellular energy supply, and by intracellular nitrate consumption, and whose activity depends on the proton electrochemical gradient. The latter system is regarded as an (H+) anion co-transport carrier mechanism that produces transient plasma membrane depolarization upon addition of nitrate. The depolarization is counteracted by the plasma membrane H+- ATP-ase. The plasma membrane proton ATP-ase is induced by nitrate.

In addition to the nitrate uptake system, plants have an inducible nitrate efflux system, requiring both RNA and protein synthesis. The efflux system, however, has a much slower turnover rate than the uptake system.

Movement of nitrite from the cytosol to the chloroplast appears to involve a proton-linked transport of nitrite. The rapid movement of HNO_2 across the chloroplast inner envelope would require some active proton transport from the stroma into the external space. The presence of an H+- ATP-ase on the chloroplast inner envelope which pumps H+ out of the stroma into the cytosol may fulfill this role.

Ammonium ions are taken up by a saturable carrier system with high substrate affinity. The low-affinity component of transport is stimulated by high external pH and probably reflects diffusion of uncharged NH_3

across the lipid phase of the plasmalemma. If ammonium enters the cell at high rates, it causes strong membrane depolarization and will block anion/H+ co-transport. This may be the primary mechanism by which ammonium ions inhibit nitrate uptake. Alternatively competition between the nitrate uptake system and glutamine synthetase for ATP may in part account for ammonium inhibition of nitrate uptake.

Role of aminotransferases in ammonia assimilation

Nitrogen, following its initial assimilation into glutamine and glutamate can be distributed to many other compounds by the action of aminotransferases. The final step in the synthesis of several amino acids is a transamination of the 2-keto analogue of the amino acid. Glutamate is often the amino-donor substrate in biosynthetic reactions, with the result that 2-oxoglutarate will be a product.

2-Oxoglutarate is a substrate required for nitrogen assimilation by most of the amino acid synthetic pathways. Glutamate-utilizing transaminases therefore regenerate the carbon precursor needed for ammonia assimilation. Regeneration of 2-oxoglutarate may permit assimilation to continue in the absence of net 2-oxoglutarate synthesis, as long as glutamate is not utilized by other reactions.

Nitrate and nitrite reductase

Ferredoxin-nitrite reductase (NiR) catalyzes the six-electron reduction of nitrite to ammonia, using reduced ferredoxin (Fd) as the electron donor. The Fd-dependent NiRs of plants, algae and cyanobacteria are monomeric proteins with molecular masses near 63 kDa which contain a single [4Fe-4S] cluster and a single siroheme (which serves as the binding site for nitrite) as prosthetic groups. Fd-NiR has been cloned from several higher plant sources, including *Betula verrucosa*, spinach and maize. The plant enzyme is distinct from the NAD(P)H-nitrite reductase of bacteria and fungi.

N demand and nitrate uptake

During vegetative growth, nitrate ions are rapidly absorbed by the root and transported via the xylem to the leaf. In the leaf, nitrate reduction produces organic acids and amino acids . Most of the newly formed organic acids are translocated to the root where a carboxyl group is released in exchange for a nitrate ion, whereas the newly assimilated N is incorporated primarily into leaf N compounds (protein). During rapid pod filling stages leaf proteolysis occurs and much of the amino N in the leaf is exported to the filling pods. Consequently the phloem is enriched with amino compounds, which repress nitrate uptake and hence diminish the rate of nitrate reduction.

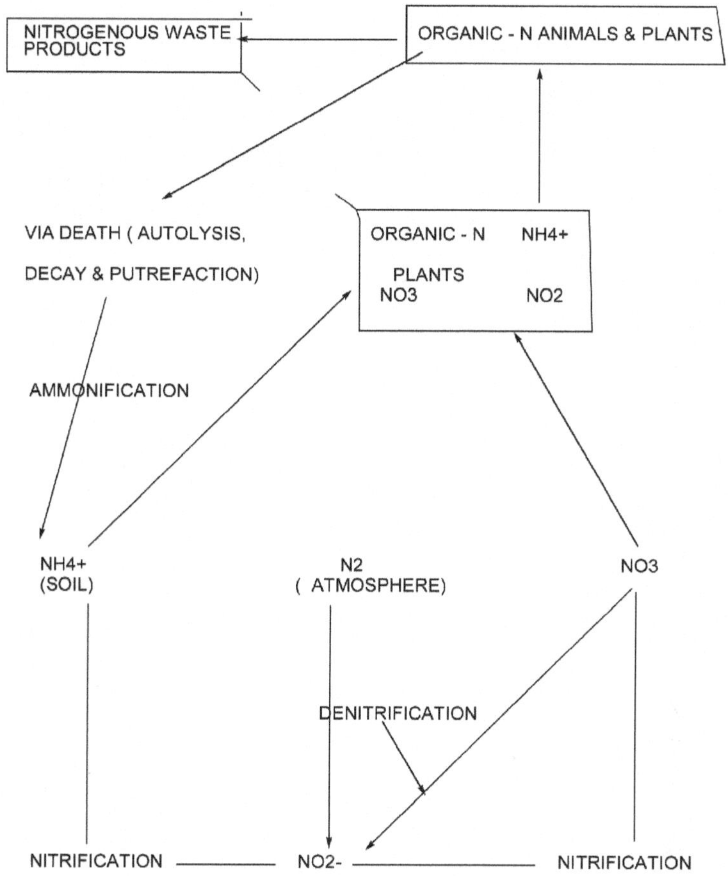

BIOLOGICAL NITROGEN CYCLE

Chapter – 22

Plant Cell

The word cell was coined by Robert Hooke with the help of compound microscope. Cell is the basic unit of life. A plant cell has three distinct components a). Cell wall b). Protoplasm and c). Vacuole. Cell wall and vacuole are considered as non-living substances. The protoplasm which is living is composed of 1. Cytoplasm and 2. Nucleus. The cytoplasm contains several organelles such as mitochondria, chloroplast, ribosomes, endoplasmic reticulum, golgi complex, lysosomes, plastids etc.

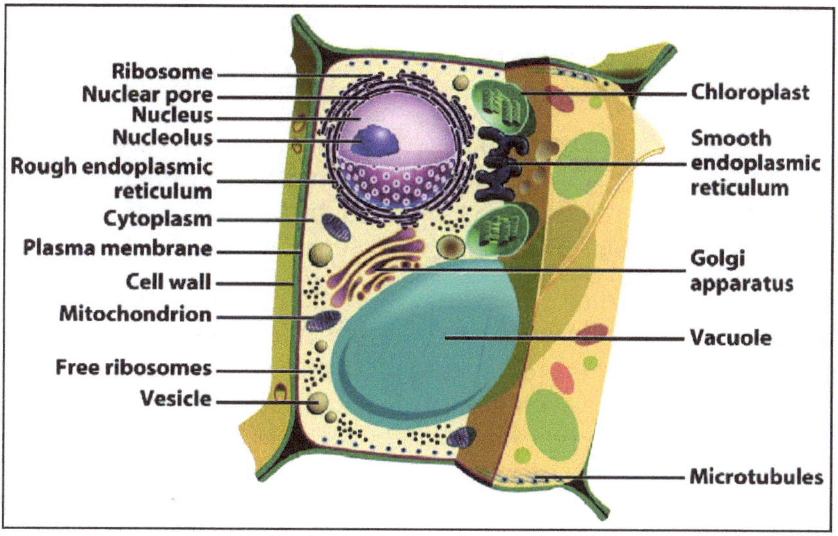

Plant Cell and organelles

Cell wall is an important component within which many organelles exist. Cytoplasm is the main component in which a nucleus and all other organelles are present. The importance, structure and functions of different parts of the plant cell are discussed below.

Cell wall

Cell wall is a non-living component of the cell and is secreted and maintained by the living portion of the cell, called protoplasm. A typical cell wall is composed of three different regions. 1. Middle Lamella, 2. Primary cell wall (1-3 μm thick and elastic), and 3. Secondary cell wall (5-10 μm thick and rigid).

Functions of cell wall: The various functions of the plant cell are as follows: 1. It protects the inner contents of the cell, 2. It gives definite shape to the cell, 3. It provides mechanical support to the tissues and act as a skeletal framework of plants, 4. It helps in transport of substances between two cells and 5. The cell wall is hydrophilic in nature and it imbibes water and helps in the movement of water and solutes towards protoplasm. It also acts as a permeable structure during absorption of minerals and solutes.

Protoplasm

It is the living, colloidal and semi fluid substance. It is also known as cytoplasm. Cell without the cell wall is called protoplast. Protoplast is enclosed by a membrane called cell membrane or plasma membrane.

Cell membrane : All cells are enclosed by a thin, membrane called plasma membrane or plasmalemma. The plasma membrane and sub-cellular membranes are collectively called as biological membrane. Cell membrane consists of proteins, lipids and other substances.

1. ***Proteins:*** The proteins present in the membranes are of two types, a. Intrinsic proteins or integral proteins which are embedded or buried in the lipid layer. These proteins associate with hydrophobic interactions to the tails of fatty acid chains of the lipid layer. In addition, integral proteins also possess hydrophilic amino acid residues which are exposed at the surface of the membrane. These proteins cannot be removed easily. b. Extrinsic proteins or peripheral proteins: They are attached to the membrane surface by weak ionic interactions. These proteins are not much involved in the architecture of the membrane. Peripheral

proteins are bound to hydrophilic proteins of the integral proteins protruding from the lipid layer.

2. *Lipids*: The cell membrane consists of phospholipids and glycolipids. The fatty acid chains in phospholipids and glycolipids usually contain 16- 20 even numbered carbon atoms. Fatty acids may be saturated or unsaturated.

3. *Other substances:* Organic compounds like polysaccharide, salicylic acid etc. are found attached to the proteins or lipids on the membrane.

Functions of cell membrane: 1. The cell membrane surrounds the protoplasm of the cell, separating the intracellular components from the extracellular environment. 2. It anchors the cytoskeleton to provide shape to the cell, and in attachment to the extracellular matrix. 3. The plasma membrane is differentially permeable and able to regulate the transport across the membrane and 4. The cell membranes maintain the cell potential.

Cell nucleus

It is oval or spherical in shape and is generally larger in active cells than in resting cells. A nucleus consists of three main parts viz. nuclear envelope, nucleolus and chromatin. The nucleus is separated from the cytoplasm by a double membrane called the nuclear envelope. The space between the outer and inner membrane is known as nuclear pore which provides direct connection between nucleus and cytoplasm. Nucleolus is a spherical, colloidal body found in the nucleus and is the place where almost all DNA replication and RNA synthesis occur. Chromatin is the basic unit of chromosome and contains genes which play important role in the inheritance of characters to offspring from parents.

Functions of cell nucleus: 1. It regulates growth and reproduction of cells. 2. The nuclear envelope allows the nucleus to control its contents, and separate them from the rest of the cytoplasm where necessary. 3. The DNA replication, transcription and post transcriptional modification occur in the nucleus.

Chloroplast

Chloroplasts are organelles found in plant cells and other eukaryotic organisms that perform photosynthesis because of the presence of the green pigment, namely, chlorophyll. They are flattened discs usually

2-10 μm in diameter and 1μm thick. The chloroplast is surrounded by a double layered membrane. The space between these two layers is called inter-membrane space. Stroma is the aqueous fluid found inside the chloroplast. The stroma contains the machinery required for carbon fixation, circular DNA, ribosomes etc. Within the stroma the stacks of thylakoids are arranged as stacks called grana. A thylakoid has a flattened disc shape and has a lumen or thylakoid space. The light reactions occur on the thylakoid membrane.

Functions of chloroplast: 1. The important processes of photosynthesis i.e, light and dark reactions occur within the chloroplast. 2. The granum is the site of NADP reduction forming NADPH+ H^+ and photophosphorylation i.e., formation of ATP in presence of light. Thus, light reaction of photosynthesis takes place in the granum region. 3. The stroma is the main site for the dark reaction of photosynthesis and 4. The chloroplast has its own genetic system and is self replicating and hence , associated with cytoplasmic inheritance.

Mitochondria

Mitochondria are rod shaped cytoplasmic organelles, which are main sites of cellular respiration. Mitochondrion is the power house of the cell. Each mitochondrion is enclosed by two concentric unit membranes comprising of an outer membrane and an inner membrane. The space between the two membranes is called perimitochondrial space. The inner membrane has a series of infoldings known as cristae. The inner space enclosed by cristae is filled by a relatively dense material known as matrix. The matrix is generally homogeneous, but may rarely show finely filamentous or fibrous structures. The matrix contains several copies of round or circular DNA molecules.

Functions of mitochondria: 1. ATP, the readily available form of energy is produced in mitochondria. 2. Krebs cycle takes place in the matrix of mitochondria. 3. The enzymes of electron transport chain are found in the inner membrane or cristae of mitochondria. 4. Heme synthesis occurs in mitochondria and 5. Controls the cytoplasmic Ca^{++} concentration.

Ribosomes

Chemically, ribosomes are ribonucloprotein complexes. Ribosomes are of two types. Ribosomes of prokaryotes have sedimentation coefficient of 70 S and consist of two sub units of unequal sizes 50S and 30 S subunits. Ribosomes of eukaryotes have 80 S sedimentation coefficient

(40S and 60 S). Two or more ribosomes become connected by a single m RNA and are called polyribosomes. The major function of the smaller subunit of ribosome is to provide proper site for binding of mRNA and its translation. The larger subunit of ribosome supports translation and translocation processes coupled with polypeptide synthesis.

Functions of Ribosomes: 1. They provide the platform for protein synthesis 2. They have the machinery for protein synthesis.

Golgi complex

Golgi bodies are an assemblage of flat lying cisternae one above the other in close parallel array. Each golgi complex has 3 to 12 interconnected cisternae which are composed of lipoproteins.

Functions of Golgi complex: 1. It helps in packaging of proteins for exporting them. 2. It plays a role in sorting of proteins for incorporation into organelles. 3. It is involved in the formation of the cell wall of plant cells.

Endoplasmic reticulum

Endoplasmic reticulum arises from the outer membrane of the nucleus forming an intermediate meshed network. It is of two types. The granular or rough endoplasmic reticulum in which the outer surface of endoplasmic reticulum is studded with ribosome and a granular or smooth endoplasmic reticulum in which the ribosomes are not attached.

Functions of Endoplasmic reticulum: 1. Rough endoplasmic reticulum is associated with the synthesis of proteins. 2. Smooth endoplasmic reticulum is associated with synthesis of lipids and glycogen. 3. It acts as an inter-cellular transport system for various substances. 4. It contains many enzymes which perform various synthetic and metabolic activities.

Vacuole

It is a membrane bound organelle found in plant cell and occupies most of the area in the plant cell. A vacuole is surrounded by a membrane called tonoplast. It is an enclosed compartment filled with water containing inorganic and organic molecules including enzymes in solution.

Functions of vacuole: 1. Isolating materials that might be harmful or a threat to the cell. 2. Stores waste products. 3. Maintains internal hydrostatic pressure or turgor within the cell. 4. Maintains an acidic internal pH. 5. Exports unwanted substances from the cell. 6. Allows plants to support structures such as leaves and flowers due to the pressure of the central vacuole. 7. Most plants store chemicals in the vacuole that react with the chemicals in the cytosol. 8. In seeds, stored proteins needed for germination are kept in protein bodies which are a modified vacuole.

Microbodies

Microbodies are ubiquitous organelles found in the majority of eukaryotic plant cells. They are mostly spherical and have a diameter ranging from 0.2μm to 1.5μm. Two types of microbodies, peroxisomes and glyoxysomes, have been characterized. These organelles differ in their distribution and enzyme composition, although both have the capacity to transform non-carbohydrate material into carbohydrate.

Peroxisomes

Peroxisomes are found in leaves of higher plants.

Functions of Peroxisomes: Peroxisomes act in parallel with chloroplast in higher plants and are believed to undertake photorespiration.

Glyoxysomes

Glyoxysomes are temporary in that they occur during transient periods in the life cycle of a plant such as in certain beans and nuts which store fats in their seeds as energy reserves. Glyoxysomes appear in the first few days after seed germination in endosperm cells and associate closely with lipid bodies. They disappear after the storage fats are broken down and converted into carbohydrate.

Functions of Glyoxysomes: Glyoxysomes are involved in the formation of sugars by the breakdown of fatty acids in germinating seeds.

Cytoskeleton

The cytoskeleton is a scaffolding contained within the cytoplasm and is made out of protein. The cytoskeleton is present in all cells. The cytoskeleton provides the cell with structure and shape. It is composed of three main kinds of cytoskeleton filaments. 1. Microfilament: They are composed of actin subunits. 2. Intermediary filaments: They function in the maintenance of cell shape by bearing tension. They also participate in the cell-cell and cell matrix junctions. 3. Microtubules: - They are like hollow cylinders mostly comprising of 13 protofilaments which in turn are α and β tubulin. They are commonly organized by the centrosome.

Functions of cytoskeleton : 1. Provides mechanical support 2. Anchors organelles 3. Helps to move substances intra cellular.

Plant cell wall

Plant cells are surrounded by a rigid, semi-permeable cell wall. The cell wall is comprised of mainly polysaccharides with some proteins and lipids. The three main polysaccharide components of the cell wall are cellulose, hemicellulose and pectin. Two types of proteins like expansin and extensin are present predominantly. Since cell wall also contains enzymes, it can be regarded as an organelle. Cell walls provide rigidity and mechanical strength, maintain cell shape and the direction of cell growth. The cell wall also prevents expansion when water enters the cell. Under the light microscope the walls separating the cells in a plant tissue are usually clearly visible. The walls of adjacent cells meet at a dividing line known as the middle lamella, which can be distinguished with the electron microscope.

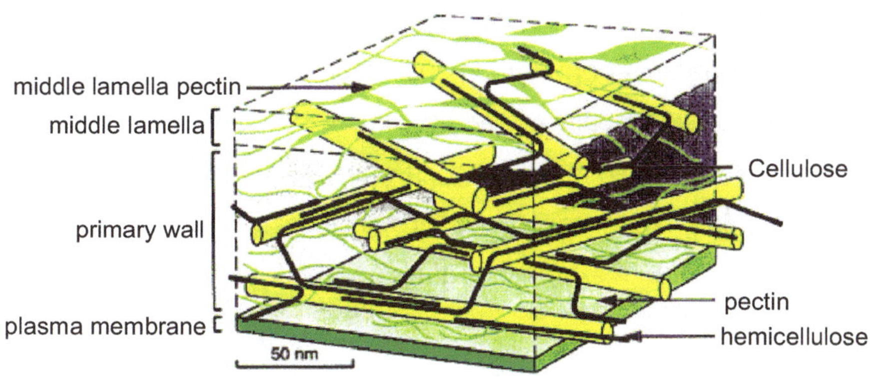

Composition of cell wall

The cell wall performs certain specialized functions. They are 1. Structural function. 2. Affects developmental pattern. 3. Defines cell's position within the plant. 4. Cell- cell and cell- nucleus communication. 5. Defense against pathogens. 6. Recognises symbiotic nitrogen fixing bacteria. 7. Recognises self . Cell wall is made up of three layers. They are:

a) **Middle lamella:** This is the first layer formed during cell division. It makes up the outer wall of the cell. It is shared by adjacent cells. It is composed mainly of pectic compounds and proteins

b) **Primary wall:** Primary wall is deposited by cells before and during active growth. Plant cells are surrounded by a polysaccharide rich primary wall. The primary walls of different plant cells differ greatly in appearance. Young cells have a very thin cell wall.

Composition of primary wall: Primary walls are composed predominantly of polysaccharides together with lesser amounts of structural glycoproteins, phenolic esters, enzymes, and calcium and boron minerals.

Functions of primary wall

- It gives structural and mechanical support
- It maintains and determines cell shape
- It resists internal turgor pressure of cell
- Controls rate and direction of growth
- It is ultimately responsible for plant architecture and form
- Regulates diffusion of material through the apoplast
- Protects against pathogens
- Protects against dehydration and other environmental factors
- Source of biologically active signaling molecules
- Plays a major role in cell – cell interaction
- Participates in early recognition of symbiotic nitrogen fixing bacteria.

c) **Secondary cell wall:** Secondary cell wall is formed after cell enlargement is completed. Some cells deposit additional layers inside the primary wall. This occurs after growth stops or when the cells begin to differentiate (specializes). The secondary wall is mainly for support and is comprised primarily of cellulose, hemicellulose and lignin. It often can be distinguished into three distinct layers, namely, S1, S2 and S3 - which differ in the orientation, or direction, of the cellulose microfibrils. It is extremely rigid and provides compression strength.

Composition of secondary cell wall: The secondary cell walls are much thicker than the primary walls and consist of 40-45% cellulose, 15 – 35 % hemicellulose, 15 – 30 % lignin and negligible amounts of pectic polysaccharides.

Functions of secondary cell wall

- It plays a major role in providing mechanical support.
- It facilitates the transport of water and nutrients.
- It allows extensive upright growth

Formation of cell wall

A cell plate formed between the two daughter cells originates from microtubules, which act as the base for the construction of the new cell wall. The microtubules may direct the cell- wall forming materials to the cell plate which grows from the centre towards periphery of the cell and soon becomes the pectin rich middle lamella. Above the middle lamella the primary wall formation takes place. The protoplasts of the cells of the primary wall secrete the secondary wall materials, when the cells have stopped enlarging. The protoplast then totally diminishes and only the wall remains. The rings, spirals or network in a mature stem cross section are due to secondary wall deposition.

Chemical changes in cell wall

Chemical changes take place in cell wall by accumulation of various depositions as the cell matures. These chemical changes bring corresponding change in the structure and function of the cell. The various depositions are:

Lignification: The deposition of the lignin on cell walls is called lignification. As a result of lignification, the cell walls become hard, thick walled and dead. Usually after of the thickening of the cell wall, the protoplasm of the cell diminishes in size and the cell becomes dead and rigid. Cellulose microfibrils are impregnated in phenolic polymer called lignin. Lignin displaces water in matrix and form hydrophobic meshwork that bonds tightly to cellulose and prevent wall enlargement. Lignins add up mechanical strength and reduce susceptibility of wall to attack by pathogens.

Suberization: The cell wall of cork cells and casparian strips of endodermis get deposited with a layer of suberin by a process known as suberization.

Cutinization: In some cells, in the outer layers of the cell wall, the cellulose gets converted to cutin by the processs of cutinization. This forms a definite, impermeable membrane on the cell wall of cuticle.

Cutinization helps in checking evaporation of water. It is found normally on the exposed parts of the plant.

Mucilaginous changes: Sometimes the cellulose is changed to mucilage which has the property of absorbing and retaining water. This forms a viscous, mucilaginous coating on the cell walls and helps to tide over dry conditions. Many sea weeds yield mucilaginous substances such as agar, alginic acid and carageenen of great commercial value.

Mineralization: The deposition of the minerals on the cell walls by the process of infiltration or by deposition of inorganic salts is known as mineralization. The minerals usually deposited are silica, carbonates and oxalates of calcium. The deposition of silica is known as silicification and calcium deposition is called calcification .

Role of plant cell wall in livestock, food and paper industry

- Primary walls are the major textural component of plant-derived foods.
- Plant-derived beverages often contain significant amounts of wall polysaccharides. Some wall polysaccharides bind heavy metals, stimulate the immune system or regulate serum cholesterol.
- Wall polysaccharides are used commercially as gums, gels and stabilizers.
- Cell wall structure and organization is of interest to the plant scientist, the food processing industry and the nutritionist.
- Cellulose plays a major role in paper industry.
- Secondary walls also have a major impact on human life, as they are a major component of wood
- It is a source of nutrition for livestock.
- The cell walls of fruits and vegetables are now recognized as important dietary components and may protect against cancer of colon, coronary heart disease, diabetes, and other ailments.
- Nevertheless, numerous technical challenges must be overcome to enable the efficient utilization of secondary walls for energy production and for agriculture.

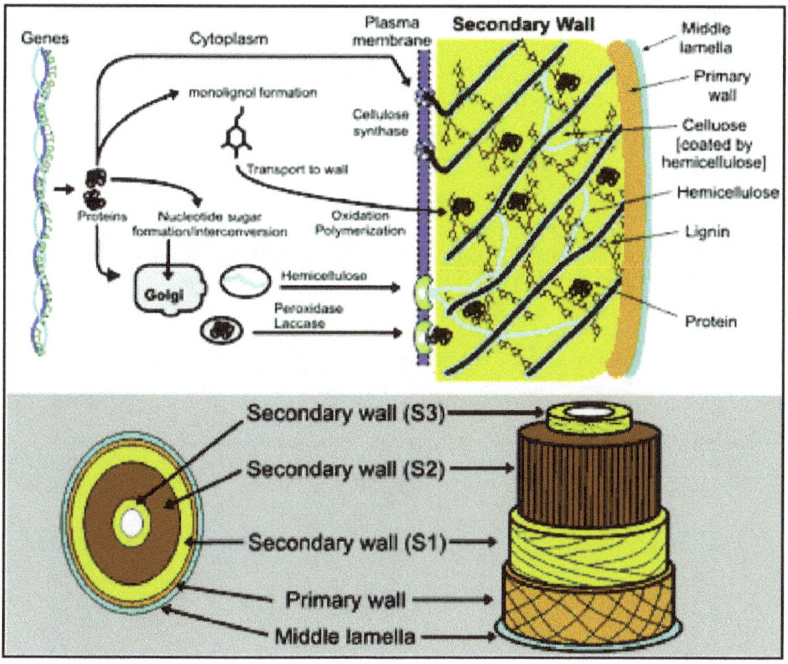

Layers of cell wall

Appendix

Conversion factors

To obtain	Multiply	By
Gram	Pounds	453.5924
Grams (cm) (sec)	Centipoises	0.01
Grams/cu cm	Lb/cu ft	0.016018
Grams /cu cm	Lb/cu in	27.680
Grams/cu cm	Lb/gal	0.119826
Inches	Centimetres	0.3937
Inches of Hg @ 32^0F	Atmospheres	29.9213
Inches of Hg @ 32^0F	Lb/sq in	2.0360
Inches/deg F	Cm/deg C	0.21872
Kg	Pounds	0.45359
Kg/cumeter	Lb/cu ft	16.018
Kg/(hr) (meter)	Centiposises	3.60
Kg/meter	Lb/ft	1.488
Kg/sq cm	Lb/sq in	0.073
Kg/sq meter	Lb/sq ft	4.8824
Liters	Cu ft	28.316
Liters	Gal	4.546
Liters/kg	Cu ft/lb	62.42621
Liters/min	Cu ft/sec	1698.963
Liters/sec	Cu ft/min	0.47193
Liters/sec	Gal/min/cu cm	0.063088
Meters	Feet	0.3048
Meters/sec	Ft/(sec)	0.3048
Pounds (avoir)	Kg	2.2046
Ponds/cu ft	Grams/cu cm	62.428
Pounds/cu ft	Pounds/gal	7.48
Pounds/ cu in.	Gram/cu cm	0.036127
Pounds/(hr) (ft)	Centioses	2.42
Pounds/inches	Grams/cm	0.0056

Contd.

Contd.

To obtain	Multiply	By
Pounds/(sec) (ft)	Centipoises	0.000672
Pounds/Gal	Pounds/cu in	231
Sq centimetres	Sq. ft	929.0
Sq centimeters	Sq.in	6.4516
Sq ft.	Sq. meters	10.764
Sq in.	Sq. centimeters	0.155
Sq meters	St ft	0.0929
Atmospheres	In Hg @ 32^{0}C	0.033421
Centimeters	Inches	2.540
Cm of Hg @ 0 deg C	Atmospheres	76.0
Cm/(sec) (sec)	Gravity	980.665
Centipoises	Centistokes	Density
Centistokes	Centipoises	1/density
Cu cm	Cu ft	28,317
Cu cm	Cu in.	16.387
Cu cm	Gal (USA, liq.)	3785.43
Cu cm	Liters	000.03
Cu cm	Quails (USA, liq.)	946.358
Cu cm/sec	Cu ft/min	472.0
Cu ft	Cu meters	35.314
Cu ft	Gal (USA, liq.)	0.13368
Cu ft	Liters	0.03532
Cu ft/ min	Cu meters/sec	2118.9
Cu ft/min	Gal (USA) liq./sec	8.0192
Cu ft/sce	Liters/min	0.0022280
Cu ft/sec	Gal/min	0.0005886
Cu in	Cu centimeters	0.061023
Cu in	Liters	61.03
Cumeters	Liters	0.001000028
Cu meters/hr	Gal/min	0.22712
Cu meters/kg	Cu ft/ lb	0.062428
Cu meters min	Cu ft/min	0.02832
Cu meters/sec	Gal/min	0.000063088
Feet	Meters	3.281
Ft/min	Cm/sec	1.9685
Ft/sec	Meters/sec	3.2808
Ft/(sec) (sec)	Gravity (sea	32.174

Summary of SI Units and convenient Metric Equivalents

Dimension	SI Base or Derived unit	Convenient SI or metric unit	SI Equivalent of Metric units
Time	Second (s) hour(h), day (d)	Minute (min) year (a)	
Length (depth)	meter (m)		
Area	m^2	hectare (ha)	$10^4 m^2$
Population, stocking Density	m^{-2}	ha^{-1}	$10^{-4} m^{-2}$
Mass	Kilogram (kg)	Tonne (t)	10^3 kg
Amount of substance mole (mole) (number of entities)			
Mass per unit area	kgm^{-2}	kg ha^{-1}, t ha^{-1}	10^{-1} gm^{-2} 10^{-2}kg m^{-2}
Mass per unit area and time	kg m^{-2} s^{-1}	g m^{-2} d^{-1}	
Volume	m^3	liter (L)	$10^{-3} m^{-3}$
Density, concentration	kg m^{-3} mol m^{-3}	mol L^{-1}	$10^{-3} m^3$
Force	Newton (N)		
Pressure	pascal (pa		
Temperature	Kelvin (K),Celsius (°C)		
Energy	Joule (J)		
Energy /unit Area	J m^{-2}	MJ ha^{-1}	$10^2 Jm^{-2}$
Power	watt (W)	MJ d^{-1}	
Power per unit area	W m^{-2}	MJ m^{-2} d^{-1}	
Photon flux density	mol m^{-2} s^{-1}	mol m^{-2} d^{-1}.	

Prefixes for multiples and sub-multiples of units

Prefix	Abreviation	Relative size of unit	Example
Nano	n	10^{-9}	ns nanaosecond
Micro	µ	10^{-6}	µg microgram
Milli	m	10^{-3}	mL millilitre
Centi	c	10^{-2}	cm^3 cubic centimetere
Deci	d	10^{-1}	dm decimetre
Deca	da	10	dam decametre
Hecta	h	10^2	hm^2 square hectometre
Kilo	k	10^3	kw kilowatt
Mega	M	10^6	Mpa megapascal
Giga	G	10^9	GJ gigajoule
Tera	T	10^{12}	Tg tetragram

Pressure Conversion Chart

To obtain	Units of Pressure					
Units x factor	ATM	kg/cm^2	psi	mm.Hg.	bars	in.H_2O
1 atm = 760mm Hg	1	1.033	14.70	760	1.013	407.1
1 kg/cm	0.9678	1	14.22	735.6	0.9807	394.1
1 psi	0.06805	0.07031	1	51.71	0.6895	27.70
1 torr =1/760 atm= 1mm Hg.	$1.316 x 10^{-3}$	$1.360 x 10^{-3}$	$19.34 x 10^{-3}$	1	$1.333 x 10^{-3}$	0.5357
1 bar=10^6 dyne/cm^2	0.9869	1.020	14.50	750.1	1	401.8
1 inch of H_2O at 15^0C	$2.458 x 10^{-3}$	$2.538 x 10^{-3}$	0.0361	1.867	$2.489 x 10^{-3}$	1

* $1F^0 = 9/5^0C + 32$

* $^0K = 0^0C + 273.16$

Concentrations-conversion table

Proportion ng/kg pg/g	potency	%	g/kg mg/g	ppm mg/kg ng/g ng/mg	ppb pg/kg pg/g pg/mg
1:100	$^{-2}$	1	10	10000	
1:1000	1x 10^{-3}	0.1	1	1000	
1:10000	1x 10^{-4}	0.01	0.1	100	
1:100000	1x 10^{-5}	0.001	0.01	10	
1:1000000	1x 10^{-6}	0.0001	0.001	1	1000
1:10000000	1x 10^{-7}	0.00001	0.0001	0.1	100
1:100000000	1x 10^{-8}	0.000001	0.00001	0.01	10
1:1000000000	1x 10^{-9}	0.0000001	0.000001	0.001	1
1:10000000000	1x 10^{-10}				0.1
1:100000000000	1x 10^{-11}				0.01
1:1000000000000	1x 10^{-12}				0.001

US, British and metric equivalents

Volume Liters	x	0.2642	=	US gallons
US gallons	x	3.785	=	Liters
Imperial gallons	x	1.201	=	US gallons
Imperiol gallons	x	4.546	=	Liters
US gallons	x	0.8327	=	Imperial gallons
Cubic meters	x	35.31	=	Cubic feet
Cubic feet	x	0.0283	=	Cubic meters
Cubic meters	x	264.2	=	US gallons
US gallons	x	0.0039	=	Cubic meters
Metres	x	3.281	=	Feet
Feet	x	0.3048	=	Meters
Metres	x	39.37	=	Inches
Inches	x	0.0254	=	Meters
Centimeters	x	0.3937	=	Inches
Inches	x	2.540	=	Centimeters
Millimeters	x	0.0394	=	Inches
Inches	x	25.4	=	Millimeters
Kilograms	x	2.2046	=	Lbs.
Lbs.	x	0.4536	=	Kilograms
Tons (long)	x	1016.05	=	Kilograms
Tons (long)	x	2240	=	Lbs.
Tones (metr.)	x	1000	=	Kilograms
Tones (metr.)	x	2204.6	=	Lbs.
Tons (short)	x	907.185	=	Kilograms
Tons (short)	x	2000	=	Lbs.
Grams	x	15.43265	=	Grains
Grams	x	0.0647989	=	Grams
Grams	x	0.0352740	=	Ounces (US)
Ounces (US)	x	28.349527	=	Grams

Energy dimensions-conversion factors

Given Required dimension with conversion factor (1)Dimension

Unit	J	kWh	MeV	mkp	$kcal_{15}^{o}$	erg
1 J	1	$2.77778.10^{-7}$	$6.242.10^{12}$	0.1019716	$2.38920.10^{-4}$	10^{7}
1 kWh	3600000	1	$2.247.10^{19}$	367097.8	860.11	$3.6.10^{13}$
1 MeV	$1.602.10^{-13}$	$4.45.10^{-20}$	1	$1.634.10^{-14}$	$3.827.10^{-17}$	$1.602.10^{-4}$
1 mkp	9.80665	$2.72407.10^{-6}$	$6124\text{-}10^{13}$	1	$2.34301.10^{-3}$	$9.806665.10^{7}$
1 $kcal_{15}^{o}$	4185.5	$1.16264.10^{-3}$	2.613.1016	426.80	1	$4.1855.10^{10}$
1 erg	10^{-7}	$2.77778.10^{-14}$	$6.242.10^{5}$	$0.1019716.0^{-7}$	$2.38920.10^{-11}$	1

(1) Examples: 1 J = $2.38920.10^{-4}$ kcal 1 MeV = $1.602.10^{-13}$ J

Estimated energy and fat requirements/day

Occupation	Work level	Energy requirement (Kcal)	Projected fat intake (g) 40%	20%
Male clerk (65kg)	Light Activity	2580	115	58
Subsistence farmer (58 kg)	Moderate activity	2780	124	62
Male heavy worker 65kg	Heavy activity	3490	155	78
Retired male (60kg)	Light activity	1960	87	43
House wife (55 kg)	Light activity	1990	88	44
Female Dev. Country —	Light activity	2235	99	50

FAO / WHO Recommended daily intake of energy

Adult man (65 kg)	Per day	per kg/day
Energy (Kcal)	3000	46
(MJ)	12.55	0.19
Protein (Egg) (g)	37	0.57
(Animal source/ mixed)	46	0.71
Cereal diet (g)	62	0.95
Iron (mg)	5	
Calcium (mg)	400-500	
Vitamin A (μg)	750	
Vitamin D (μg)	2.5	
Thiamine (mg)	1.2	
Reboflavin (mg)	1.8	
Niacin (mg)	19.8	
Folic acid (μg)	200	
Vitamin B12 (μg)	2.2	
Ascorbic acid (mg)	30.0	

Vitamin Requirement

Vitamin	Qnt	Children 0.5/4-6 yrs.		Males 11-14; 23-50		Females (Pregnant)
A	µg	420	500	1000		
	6 1000	1000				
D	µg	10	10	10	5	10
E	mg	3	6	8	10	10
K	µg	12	20-40	50-100	70-140	70-140
Thiamine	mg	0.2	0.9	1.4	1.4	1.5
Riboflavine	mg	0.4	1.0	1.6	1.6	1.6
Niacin	mg	6	11	18	18	18
Vit. B6	mg	0.3	1.3	1.8	2.2	2.6
Pantothenic	mg	2	3-4	4-7	4-7	4-7
Biotin	µg	35	85	100-200	100-200	100-200
Folic acid	µg	30	200	400	400	800
B12	µg	0.5	2.5	3.0	3.0	4.0
C	mg	35	45	60	60	80

Mineral element requirement

Element	Infants	Children	Males		
	0.5-1.0 yr	4-6`	15-18	25-30	51+
Sodium mg	500	900	1800	2200	2200
Calcium mg	540	800	1200	800	800
Potassium mg	1100	1550	3250	3900	3900
Phosphorous mg	360	800	1200	800	800
Magnesium mg	70	200	400	350	350
Copper mg	0.85	1.7	2.5	2.5	2.5
Iron mg	15	10	18	10	10
Zinc mg	5	10	15	15	15
Manganese	0.85	1.75	3.75	3.75	3.75
Iodine µg	50	90	150	150	150
Selenium mg	0.02-0	0.03-0.12	0.05-0.2	0.05-0.2	0.05-0.2
Molybdenum mg	0.04-0.08	0.06-0.15	0.15-0.5	0.15-0.5	0.15-0.5

FSC
www.fsc.org
MIX
Papier | Fördert gute Waldnutzung
FSC® C083411

Zeitfracht Medien GmbH
Ferdinand-Jühlke-Straße 7
99095 Erfurt, Deutschland
produktsicherheit@kolibri360.de